The Numerate Leader

The Numerate Leader

How to Pull Game-Changing Insights from Statistical Data

Thomas A. King

WILEY

Published by John Wiley & Sons, Inc., Hoboken, New Jersey.

Published simultaneously in Canada.

Library of Congress Cataloging-in-Publication Data

Names: King, Thomas A., 1960- author. | John Wiley & Sons, Inc., publisher.
Title: The numerate leader : how to pull game-changing insights from statistical data / Thomas A. King.
Description: Hoboken, New Jersey : Wiley, [2022] | Includes bibliographical references and index.
Identifiers: LCCN 2021027855 (print) | LCCN 2021027856 (ebook) | ISBN 9781119843283 (cloth) | ISBN 9781119843306 (adobe pdf) | ISBN 9781119843290 (epub)
Subjects: LCSH: Statistical literacy. | Commercial statistics.
Classification: LCC QA276 .K483 2021 (print) | LCC QA276 (ebook) | DDC 001.4/22—dc23
LC record available at https://lccn.loc.gov/2021027855
LC ebook record available at https://lccn.loc.gov/2021027856

Cover Image: Edmond Halley: ©GeorgiosArt/Getty Images
Comet: ©ZU_09/Getty Images
Postage Stamp: ©IkonStudio/Getty Images
Cover Design: Wiley

SKY10029753_091321

CONTENTS

PREFACE

Since you've cracked open this book, permit me a few guesses. In your educational journey, you have taken at least one introductory statistics class. That course involved associating types of story problems with many formulas. Success meant matching formulas to stories and then plugging and chugging to arrive at so-called answers that could be measured to decimal places.

You then prepared for a final exam by cramming facts into short-term memory. You arrived at the testing location at the appointed hour, attacked assigned questions, submitted the completed test, celebrated with friends, and then cleared your mind to gird for the next task on that semester's to-do list. Today, any effort to recall lessons learned yields little beyond hazy memories.

This story played out four times in my academic travels. I studied introductory statistics in high school, college, a master's program designed to teach accounting to liberal arts students, and then an MBA program. Receiving decent grades from each trip to the well, I viewed myself as quantitatively literate and accepted an insurance job with confidence. The Greeks had another word for this trait – *hubris*, outrageous arrogance, which brings misfortune to those so afflicted.

My assignment was simple: review historical car accident data to set future prices. If prices were too high, we wouldn't sell any insurance; if too low, resulting claims costs would swamp any premiums collected and bring financial loss. Should pricing actions bring profitable growth, I would be rewarded; failure, encouraged to pursue other career opportunities.

I stumbled through early assignments with a track record unblemished by success. Time spent in the classroom offered little benefit when faced with messy, real-world data. Sufficiently self-aware to recognize that I did not know what I was doing, I sought help from more informed coworkers.

The Beatles sang how they got by with a little help from their friends. Support offered by numerate colleagues salvaged my career. My mentors invested time to teach me how to think about the basic concepts discussed in this book.

I was then able to make a decent living applying these tools to reveal patterns buried in unfamiliar data sets, a skill rewarded by labor markets. Evidence of success was the ability to pay off our mortgage, send three kids to private colleges, and then secure my wife's permission to become an underpaid university professor at middle age.

More interestingly, Peter Lewis, Progressive's leader for much of my career there, set up a process to hire and train dozens of people like me. Analytic skills unleashed by this business model helped our company, one of many in a fragmented, mature industry, grow organically from obscurity into a Fortune 500 firm. In this process, Peter earned his way onto the Forbes 400, a compilation of the richest people in the world. Using this one data point, I infer that numerate organizations create wealth.

I now teach accounting to undergraduates, graduate students, and executives. Conversations with our alumni have convinced me that the power of statistical reasoning lies not in fancy tools or sophisticated software. Rather, informed decision-making results from the ability to remember and apply a few basic ideas. This book is a (mostly) qualitative introduction to these ideas.

My thesis is that anyone armed with a basic understanding of the ideas that follow may surface information from unfamiliar data sets. Turning data into information allows ordinary people to make material contributions to their employers and society.

This book is designed to raise the statistical game of a generalist who works on an organization's front line. The target reader is too busy with budget deadlines, screaming customers, supply chain problems, and human resource mishaps to learn exotic math.

To borrow a phrase from political adviser James Carville, *It's the ideas, stupid.* This book is not about which buttons to push when firing up a computer. All software discussions provided within these pages are limited to a few functions in Microsoft Excel. This is the book to read before enrolling in advanced data analytics classes.

Numerate employees who use these tools identify provisional relationships. We then invite statistical experts, whom I respectfully call propeller heads, to kick the tires and determine whether proposed ideas have merit. If experience is any guide, identifying just one or two insights that stand up to scrutiny will be a game-changer for the reader's career and their employer's financial prospects.

The core skill is finding patterns that may be expected to recur in the future. The recipe discussed in the following pages blends concepts you learned in introductory statistics laced with a few ideas from the philosophy of science. You have already learned – but likely don't yet understand – what's covered in the pages that follow. What is new is how the material is explained. My hope is that my storytelling skills will allow you to put previously learned material to much more productive use.

As a warning, I am not a math person. Statisticians reading this book will quickly tick off justifiable criticisms about incompleteness and lack of rigor. My defense is that nontechnical explanations are a necessary step to help generalists uncover interesting questions that permit propeller heads to work their magic.

Academics teach doctoral students to write in dry, serious prose. I have deliberately chosen to ignore my scholarly training and write in the first person with a conversational voice. Doing so, I hope, will make statistics less scary and spark conversations that will make the world a better place. Numerate people seek to be approximately right rather than precisely wrong.

Writing a book is more of a journey than a task. An African proverb says that if one wishes to travel quickly, then go alone; if one wishes to travel far, then go together. My journey followed the second approach, and I'm too embarrassed to share how much time went into the creation of this slim volume.

Any effort to thank all those who contributed to my journey would rival a bad Oscars' acceptance speech or, even worse, the drudgery of the *Iliad*'s catalogue of ships. Instead, I simply express gratitude to the people of The Prairie School, Twin Disc, Harvard College, New York University, Arthur Andersen, Harvard Business School, Progressive Insurance, and Case Western Reserve University. Colleagues and mentors at these organizations invested countless hours in my intellectual development.

I do wish to acknowledge my paternal grandfather, who encouraged me to write and stressed the importance of an active voice, strong verbs, and the measured use of prepositions, and my father, who taught me that there is never a single, correct answer to any significant management problem.

A special shout-out goes to my academic mentor, Gary Previts, who gave me the opportunity to sample the academic life. My students and colleagues at the Weatherhead School of Management pushed me to refine my thinking with patience and good humor. Finally, whatever success I have achieved is due to the boundless support given by wife, best friend, and life partner, Yvonne.

Responsibility for errors and omissions in the following pages rests squarely on my shoulders.

Chagrin Falls, Ohio
May 2021

1

NUMERACY

Our journey begins with the story of Edmond Halley (1656–1742). Son of a wealthy soap maker, he studied mathematics at Oxford and became a respected astronomer. The question of how gravitational force influences the shape of celestial orbits led him to travel to Trinity College, Cambridge, to seek help from the reclusive Isaac Newton.

Upon their meeting, Halley recognized the scope of Newton's astonishing genius. The young astronomer pressed his hero to publish thoughts in a book that came to be titled *Mathematical Principles of Natural Philosophy*, which has my vote for the most significant scientific book ever written. Among other things, Newton demonstrated that the sun holds a planet in an elliptical orbit.

Halley then used this principle to review past observations of comets. These bright, fleeting objects have attracted attention since biblical times. The Bayeux Tapestry shows a comet streaking through the sky before the Norman conquest of England in 1066. Halley reasoned that comets follow the same laws of motion as other celestial objects.

Armed with Newton's inverse-square law, Halley studied records of past comet observations and attempted to infer orbital periods. He concluded that a comet he observed in 1682 was the same object identified in the tapestry and then seen by Peter Apian in 1531 and Johannes Kepler in 1607. In 1705, Halley predicted that this object would return to view in 1758.

Sixteen years after Halley's death, stargazers turned to the skies to search for the returning comet. On Christmas Day in 1758, an astronomer in Germany spotted the object that would become known as Halley's Comet.

This anecdote illustrates numeracy in action. Halley identified a pattern buried within data and then used this pattern to make a bold prediction outside of the domain of the original data set. This was some trick because the gravitational pulls of Jupiter and Saturn give the comet an irregular orbital period. What's more, this object is visible for just a few weeks out of each orbit. Halley crafted a successful prediction from thin data.

Halley's story leads to the summary idea of this book:

Numeracy is the *craft* of *statistical reasoning.*

Permit me some ink to unpack the four key words in this short sentence.

Numeracy. I first came across the term *numeracy* as a college student, when I read a paper written by the statistician Andrew Ehrenberg (1977). He railed against colleagues at the prestigious Royal Statistical Society who amassed gobs of data and then did little with it. Educated math whizzes, surrounded by rich data sets, often show little skill at extracting meaningful information and then presenting it clearly. A useful statistician – for whom Ehrenberg bestows the label of numerate – identifies a cool "so what" and then presents the finding in a way that communicates it clearly.

I cannot begin to list all the horrifically bad presentations I have endured in my career. Time and again, well-trained people gushed waterfalls of data without providing any meaningful insight. Apparently, the presenters were saying, *I worked so hard on this analysis, and I want you to see evidence of how much time was spent on the project.* Ugh.

Occasionally, however, I watch a presentation or read a paper where the presenter distills hours of work into a simple summary that builds a bridge between evidence and conclusion. When I'm blessed to receive the work of a numerate person, I feel enormous gratitude for the gift of information delivered in a concise, pleasing manner.

Halley showed numeracy by distilling his work on comets into a breathtakingly simple conclusion: the celestial body in question would again be visible in 1758. Simplicity is a foundational idea of this book.

Craft. A typical U.S. university has a school of arts and sciences. Disciplines taught there, ranging from soft humanities to hard physical sciences, expose students to two broad categories of scholarship. The softer arts encourage expression of individual points of view while the sciences emphasize agreed-upon answers.

No two classics students will draw identical conclusions about the role of Achilles in the *Iliad,* but every chemistry student should agree on how to balance a reaction equation involving sodium and hydrochloric acid. Scientific statements may be disproven, while those in the humanities may be argued without end.

Falling between arts and sciences are crafts, pursuits requiring a type of thinking rarely taught in higher education. An example of a craft is metalworking. Grades of steel have chemical properties that lend themselves to scientific study. Other factors, however, influence a metalworker's ability to cut a certain piece of metal to meet required specifications.

On a particular day, the air has a certain temperature, humidity, and pressure. The grinding machine is at a certain stage in its maintenance cycle. The steel blanks have idiosyncratic properties associated with the production lot made at a point in time at a particular mill. I doubt that any scientific model will ever be able to incorporate how these and dozens of other variables influence the shaping of metal.

When I was in college, I worked in a factory that made transmission assemblies. An unshipped unit in final assembly needed roller bearings with dimensions specified to a couple ten-thousandths of an inch – a ridiculously tight tolerance in an age before the spread of numerically controlled machines. My job was to deliver steel blanks to a master machinist who was "voluntold" to cut the bearings immediately. A baseball analogy would be asking a player on his scheduled day off to step out of the dugout in the bottom of the ninth inning to hit a home run.

A crowd formed around the machinist, who examined the blanks, inspected the grinding machine, and took note of myriad factors that were lost on me. He set up his equipment and began cutting. A quality control professional used precise calipers to measure the dimensions of the output. The first few bearings failed quality control. The machinist made adjustments and then produced a series of bearings that each met the required size standards. The rest of us watched in awe.

In that moment, this craftsman garnered more respect among colleagues than any investment banker, management consultant, university professor, or business executive I've ever met. None of us present that day could come close to doing what he just did. I considered dropping out of college to become his apprentice.

This machinist – whom I remember as Yoda – worked a craft, a discipline that straddles the domains of art and science. The science of metallurgy informs us of processes used to transform steel. However, the scope of this science is not sufficiently developed to tell us how to handle every situation we may face. At that point, a craftsman blends individual judgment with formal training to accomplish a desired task.

This judgment is not easily codified or documented, hampering the ability of a master to pass along know-how to an apprentice. The master offers coaching, but the apprentice assumes responsibility for finding their own way while learning from the master.

Numeracy is a craft. There is some science embedded in the tools used to reveal patterns buried in data. However, this science is not sufficiently robust to instruct people what to do in all cases. Simply buying a computer loaded with statistical software gets one nowhere fast. Numerate people use their wits to sort through quirks embedded in unfamiliar data sets. An effective craftsman blends school-taught technique with hard-won experience to sort through the problem at hand.

The little science given in this book merely repeats what many other reference books on statistics share. Any decent teacher may explain the dozen or so concepts discussed here. The real magic comes from you using your judgment to apply them to circumstances associated with data sets in your life. The art of numeracy is learned but not taught, and I hope that the storytelling in this book serves as a catalyst to help you cultivate this skill faster than what would have been accomplished if you had not read these pages.

Halley's prediction required blending the science of Newton's inverse-square law with the art of estimating masses and distances of significant bodies within the solar system using incomplete astronomical data available at that time.

Statistics. Statistics, a subset of mathematics, studies how one may make uncertain inferences about a broader (and often unobservable) population from the study of properties of a small sample. A classic example is when your grandmother prepared homemade soup. After combining and heating the ingredients, she undoubtedly stirred the liquid and sipped a spoonful to assess the mixture. She was able to draw conclusions about the entire pot from a small taste.

Statistics may be viewed as the opposite of probability, the study of the likelihood of future events occurring based on known frequency distributions. We know that a fair, flipped coin has a 50% chance of landing as a head. Since fair coin flips are independent events, we may conclude that there is a one-in-four chance that this coin will land as heads in two consecutive trials.

The field of probability arose in the seventeenth century as gamblers sought to understand how much money should be wagered in games of chance. Use of probability theory requires that the objects studied – be they coins, cards, dice, or roulette wheels – have knowable probability distributions. This pursuit was popularized in the 2008 movie *21*, where a group of card-counting MIT students used probability theory to make money at blackjack tables at Las Vegas casinos. Unfortunately, most phenomena in our lives are more complicated than games of chance.

Frank Knight, a little-remembered economist, made this distinction by contrasting risk and uncertainty (1921). Risk refers to circumstances where things we study have known probability distributions. As shown in the movie *21*, decks of cards meet this standard. In a well-shuffled deck of 52 cards, there is a 1-in-13 chance that the next card drawn will be a queen. Peter Bernstein wrote the classic discussion of how leaders over the years have used tools of probability to bring risk under control (1996).

By contrast, uncertainty represents situations where outcomes have unknown (and perhaps unknowable) frequency distributions. Helping people make sense of uncertain situations is the contribution of this book. Whether a firm should expand into a new market is such a problem. The likelihood of outcomes may be expressed in approximate terms (*Boss, there's a decent chance that our product will catch on with French consumers*) but almost never as a precise number.

Knight argued that it is difficult to earn sizable profits from situations involving simple risks. Others may perform the same calculations and neutralize any advantage gained from applying probability tools to the problem at hand. However, the ability to tame uncertainty offers the prospect of substantial economic rewards.

Inferences arising from statistical analysis are always uncertain. Our sample may have been too small or not representative of the broader population. Users of statistical analysis must accept that conclusions reached may be completely wrong. The reward for trying, however, is that, with practice, numerate people are better positioned to reap economic profits described by Knight.

Halley demonstrated numeracy by offering a range of dates for his prediction. He did not express his prediction as a point estimate (i.e., a particular day) but instead offered a confidence interval (a year) in which his prediction was expected to be realized. To borrow a line from the movie *Love Story*, statistics means never having to say you're certain.

Reasoning. Reasoning, the final operative word in our definition, means connecting dots. A well-reasoned argument shows the bridge that links the evidence to the conclusion.

Halley connected the dots by combining deduction (applying the general principle of Newton's inverse-square law of gravitation to the particulars of comets orbiting the sun) and induction (generalizing from specific messy, incomplete observational data) to reach a justifiable conclusion. Halley went even further by expressing his conclusion in such a way that it could be disproven. The fact that the comet returned in the predicted year does not prove that Halley was right but making a prediction that came to pass gave the guy a lot of street cred.

Numeracy is thus the ability to turn raw data into ***information.*** Claude Shannon, a polymath working at AT&T's Bell Labs, wrote one of the most significant papers of the twentieth century (1948). Bearing the sterile title "A Mathematical Theory of Communication," the article argues that information is something real that may be defined, measured, and managed.

Shannon's crucial point is that information is a surprise. If a message is predictable, then no one is surprised to receive it, and it thus contains little information. An example would be an old friend repeating an oft-told story, where you pretend to listen as the tale inches toward its predictable conclusion. In contrast, a message is informative if it is unexpected. A great murder mystery meets this standard because few readers connect the clues to solve the crime.

Imagine going to a college reunion, where former classmates cluster in groups to catch up on life events. The groups collectively produce a stable level of background noise. From time to time, this baseline is punctuated by shrieks as one group reacts to the sharing of an unexpected, juicy piece of gossip. Outbursts reveal elevated levels of information transfer. A second analogy for a surprise is the large stock price reaction to a release of significant but previously undisclosed information about a company's business prospects.

In Shannon's world, information reduces uncertainty. The purpose of studying numeracy is to extract information from raw data. To use an academic word, numeracy mediates the process of extracting information from available data.

Insight gained – uncertainty reduced – better positions us to cope with the messy world surrounding us. Using Knight's framework in a commercial setting, information reduces uncertainty and supports efforts to earn economic profits. My former employer and its CEO earned considerable monetary rewards from investments that boosted employee numeracy.

The remaining pages present the argument that unlocking the promise of numeracy comes from understanding a limited number of basic ideas, illustrated as letters, symbols, or brief formulas.

I warn you that time invested in this book is no guarantee of success. Learning a craft takes practice, but practice is not

Figure 1.1 Numeracy Helps Extract Information from Data

a sufficient condition for achieving desired goals. The essayist Malcolm Gladwell studied factors associated with extreme levels of performance across many realms of human achievement (2008). He found that raw talent is a poor predictor of success.

Instead, Gladwell believed that highly successful people are willing to invest 10,000 hours to develop a skill. Regardless of discipline – chess, music, sports, computer programming, whatever – experts put in the hours to hone their craft. A 10,000-hour investment represents spending four hours per day, five days per week, over a decade to develop a skill. The machinist I admired had made such an investment to learn his craft.

My belief, however, is that a few hours spent reading this book will accelerate the learning process, much like a catalyst speeds a chemical reaction without direct participation. Time spent with this book allows the reader to realize a greater return from their 10,000-hour journey. If I've done my job well, reading this book will allow you to make more meaningful contributions to organizations that you serve.

The good news is that data analysis becomes easier and more enjoyable with practice. I have reached the stage where it is fun to explore new data. Really.

Before closing this chapter, let me discuss what numeracy is not. First, numeracy is not the answer to all problems of management. The craft of statistical reasoning is merely additive to any effort to predict or explain things.

My beloved Cleveland Browns, an American football team, have not done well in the (many) years associated with my writing this book. As I pen these words, there are 32 teams in the National Football League, and my Browns are one of just four to have never made a Super Bowl appearance. Player turnover has been so high that I cannot name all the starting quarterbacks since the team's 1999 reincarnation. During the two seasons ending in December 2017, the Browns won just one game and lost 31 (or a winning percentage of 3.125% [1-in-32]), a rate statistically worse than flipping a coin – evidence that something was broken.

On a trip to Kansas City, I had the opportunity to meet an executive with the Kansas City Chiefs, a rival team that flourished when my Browns languished. I asked him why my team has fared so poorly. Without hesitation, he replied that we had placed excessive reliance on "Harvard stats guys." His criticism was that the Browns had given too much authority to highly educated people who put inordinate faith in statistics. Good leadership mixes art with science. Numbers are helpful but cannot provide all the answers.

Second, numeracy does not mean mathematical rigor. Any reader armed with a decent grounding in high school math is capable of understanding concepts discussed in this book. The role of the numerate person is to identify provisional patterns buried in raw data. Use of these patterns, should they stand up to scrutiny, may transform organizations. Progressive's experience in using patterns to predict insurance accidents is a notable example. Success does not require the reader to learn trendy data analytics tools, such as data mining, artificial intelligence, or machine learning, that have garnered lots of attention in the business press.

What is needed is the help of trained statisticians to kick the tires of provisional patterns surfaced by numerate generalists. After we do our work, we invite propeller heads to use skills I will never acquire. These experts, armed with advanced degrees in mathematics or computer science, articulate hypotheses, aggregate data from various sources, and then use a host of advanced techniques to rule out proposed ideas unsupported by data. Propeller heads extend the work of the numerate. Generalists and propeller heads are teammates, not rivals.

Propeller heads are trained professionals with individual identities, not unlike doctors. Physicians have received considerable education and training to offer specialized health care services. A patient with a vision problem would typically not reach out to a dermatologist for help. Numerate generalists involved in, say, refining an opinion poll would typically not ask an econometrician (one who works with large, structured data sets) to try to disprove a hypothesis.

To take the medical analogy further, a generalist knows that no two experts will agree on everything. A patient suffering from macular degeneration may receive conflicting diagnoses and prescribed treatments from two ophthalmologists. A numerate employee may similarly receive conflicting advice and conclusions from statisticians. Welcome to the game. Finding ways to work with propeller heads is part of the 10,000-hour journey to become numerate.

To sum up, this book seeks to cultivate numeracy. Whereas a literate person can read and write, a numerate person is able to count and compare. These skills better enable someone to make informed predictions or explanations that help organizations flourish in an uncertain world.

Prediction – a means of coping with danger – lies at the core of human survival. Whoever makes better predictions garners advantage over others. The ancient Greek philosopher Thales (circa 624–546 BCE) allegedly used powers of observation and reasoning to corner the market for olive presses based on weather forecasts that foretold a healthy crop yield. He also predicted a solar eclipse.

A second purpose of numeracy is to support efforts to explain why things have happened as they have. Dinosaurs are extinct. Prediction offers little help in assessing their prospects. However, an explanation for their extinction offers a useful story that better equips us to plan for environmental change.

Any student of management needs to improve his or her ability to predict or explain things. Numeracy lies at the heart of these skills. This is the book I wish someone had handed me when I began my career.

Recap of Chapter 1 (Numeracy)

- Numeracy is the craft of finding patterns buried in data sets.
- Numerate people convert raw data into uncertainty-reducing information.
- The craft of numeracy permits more informed predictions or explanations.
- The foundation of numeracy rests on a dozen or so statistical ideas presented in this book.

2

ZERO [0]

Perhaps the most underrecognized idea in human thought is the number *zero*, an idea that emerged in many places in ancient times. The signifier, an oval-shaped symbol, is recognizable to schoolchildren around the world. The signified, the concept associated with the oval, is the absence of something to be measured. The ability to show the presence of an absence makes numeracy possible.

The number zero confers at least three benefits. First, it serves as a placeholder so that we may measure differences between numbers. Second, zeroes may sometimes be placed to the right of decimal places to create ever-more-precise measures for items of interest. But, most importantly, zero allows the establishment of a baseline so that we may calculate ratios. Numerate people need to be aware of when we may not be able to calculate differences, ever-more-precise measures, or ratios.

Placeholders. Let's start with the number 3,210. The figure has a value equal to the sum of three thousands, two hundreds, one ten, and no single units $[(3 \times 1{,}000) + (2 \times 100) + (1 \times 10) + (0 \times 1)]$. If we wanted to add or subtract a similar figure to this value, we would align counts for thousands, hundreds, tens, and ones for the two figures involved and then perform the arithmetic on counts within component categories.

Children in primary school learn to align two or more numbers and then borrow or carry from adjacent categories to perform long addition and subtraction. The use of zeroes in our Arabic numeral system permits first-graders to calculate differences between large numbers.

Now, imagine we use a number system with an absence of the number zero. An example is one used in Ancient Rome. Roman numerals assign symbols for certain quantities. The letters M, D, C, X, V, and I represent, respectively, the values for 1,000, 500, 100, 10, 5, and 1.

To cope with not having zeroes, we place smaller values to the left of a larger value to signify subtraction or on the right to signify addition. Thus, IX means that the reader should subtract 1 from 10 to arrive at 9, and XI means that the reader should add 1 to 10 to arrive at 11.

To demonstrate how horribly cumbersome this approach is, consider a stylized example presented in Table 2.1. Here we show the results of a toll-taker collecting a tax from pedestrians who cross a bridge in the Roman Empire. The civil servant

Table 2.1 Pedestrian Bridge Crossings over a Week

	Cumulative Counts		*Daily Counts*	
Day	*Roman*	*Arabic*	*Roman*	*Arabic*
Sunday	IX	9	IX	9
Monday	XLIX	49	XL	40
Tuesday	XCVIII	98	XLIX	49
Wednesday	CDXCIX	499	CDI	401
Thursday	MIV	1,004	DV	505
Friday	MCXCVII	1,197	CXCIII	193
Saturday	MDVII	1,507	CCCX	310

simply tallies the number of people who cross each week. The figure grows over the week as the cumulative number of pedestrian crossings increases.

The toll-taker reports to an administrator (the tax man) who wants to know how many people cross the bridge daily in order to make granular predictions of future tax revenue.

The records show that, through Wednesday, CDXCIX people (or 499, using Arabic numerals) had crossed the bridge, and this cumulative number had grown to MIV (1,004) through Thursday. Suppose the tax man asks how many people crossed on Thursday.

Here's my response to the boss's question:

$$\text{MIV} - \text{CDXCIX} = \text{DV}$$

$$1{,}004 - 499 = 505$$

The use of zeroes as placeholders, provided in the Arabic numbering system, makes this subtraction problem child's play, whereas the absence of zeroes in the Roman system makes the problem quite difficult. Zeroes let us measure differences easily.

Precision. When placed to the right of a decimal place, zeroes help us measure things with increasing granularity. Suppose an interior designer asks us to measure the width of the family room in our house. We offer three responses:

1. 5 meters
2. 5.0 meters
3. 5.00 meters

All three measurements are equal to about 16 feet, 5 inches. Yet, each response conveys different information. A measurement made to the nearest hundredth of a meter permits more exact space planning than would be available for a measurement made to the nearest tenth. The first measure is useless if we need to buy new furniture or a rug.

However, we may not always be able to measure things with increasing precision. Let us now add three words to our vocabulary: dichotomous, discrete, and continuous. Whenever looking at a new data set, please classify observations into one of the following three buckets.

Dichotomous variables display just one of two possible values. A switch is on or off, an employee is present or absent, or a flipped coin lands as a head or tail. Yes, we may experience unusual circumstances (say, coins landing on edges), but these events are outliers – something considered in Chapter 7.

People *tally* dichotomous variables, such as the number of days an errant employee fails to show up for work at the factory. We need to be careful when selecting statistical tools to evaluate dichotomous phenomena, such as whether or not a firm declares bankruptcy within a given year.

Discrete variables may assume certain values. A rolled die displays a value of one through six. A dealt card from a standard deck must display one of four suits (heart, diamond, club, or spade), but the suit color is dichotomous (either red or black). In the United States, one may buy shoes in sizes 8, 8½, or 9 but not in, say, size $8^2/_3$. People *count* discrete variables, such as the number of completed transmission assemblies shipped by the factory in a month.

An issue with discrete variables is determining the thing to be counted. There's the joke about a cashier working in the express lane of a grocery store in Cambridge, Massachusetts. A shopper approaches with a cart packed full of groceries. The employee nods to the overhead sign warning "Twelve or fewer items" and asks the shopper whether he went to Harvard and can't count or MIT and can't read.

The joke highlights a serious issue: How do we count observations in a data set that requires words to describe the units? All cashiers would likely count a bag of 10 apples as one item. A BOGO (buy one, get one) promotion on tubes of toothpaste also probably counts as one item. If the store offers to sell five single-serving containers of yogurt for $4, should the bundled purchase be considered one item or five?

Suppose a shopper visits the meat counter and places an order for 10 pounds of hamburger and receives three wrapped packages with varying weights. Should the cashier evaluate the purchase as one item or three? How we count observations determines sample size (n), the topic of Chapter 3, which is a basis for assessing the robustness of conclusions reached in our statistical analysis.

Continuous variables may assume a wide range of values. The length of the room described earlier is an example. Foreign currency exchange rates are often quoted to several decimal places. As I write these words, a currency dealer is willing to sell you euros at $1.1829 or buy them from you for $1.1828, a tight spread suggesting the presence of a liquid foreign exchange market. People *measure* continuous variables, such as the cost of goods sold for all the transmission assemblies shipped in a month.

Some textbooks say that continuous variables may take on any value. This is not true. Software limitations or space constraints restrict the number of significant digits that may be displayed. Stock prices quoted on the New York Stock Exchange, considered to be continuous data, are measured down to pennies but not fractions of pennies. The Excel software on my computer does not store values beyond 15 significant digits. My crude rule of thumb is that any data able to be reliably measured to three significant digits may be classified as continuous.

The point of this section is that dichotomous and discrete data sets may not be evaluated beyond a given level of precision, while continuous data sets have practical limitations on measurement. As much as we love zero, there are times when we may not use it.

Ratios. The third use of zero, the establishment of a baseline from which one may calculate ratios, is not always possible. Just after World War II, psychologist Stanley Stevens published a paper that distinguished four types of scales to measure things

(1946). He propagated four more words that we must add to our vocabulary: nominal, ordinal, interval, and ratio. I remember them using the mnemonic "Film Noir."

Nominal data fall into categories that have no logical order. Examples include eye color (e.g., green, blue, brown, hazel), religion (Islam, Judaism, Christianity, Buddhism), or make of automobile (Ford, Chevrolet, Toyota, Honda). Observations may be listed alphabetically or in order of how often they occur, but any ranking requires reference to information outside of the category labels.

Ordinal data may be placed into a lowest-to-highest sequence. Examples include ranks of officers within a branch of the military or the bronze, silver, and gold medals awarded at the Olympics. The rub is that there is no easy way to measure differences between classifications. Promotion of one rank from Colonel to Brigadier General in the U.S. Army (from pay grade O-6 to O-7) is a much bigger deal than promotion from Second Lieutenant to First Lieutenant (O-1 to O-2). The general rule is that we cannot use the tool of subtraction to measure differences between classifications within an ordinal data set.

Interval data have observations that are separated by uniform differences but are not measured from an absolute baseline. Examples include time of day and degrees Fahrenheit. Interval data may be added or subtracted but not multiplied or divided: 4 p.m. is two hours later than 2 p.m. but is not twice as late; 60° is 15 degrees warmer than 45° but not one-third hotter.

Ratio data are the gold standard of scales because observations are measured in terms of equally spaced units on top of an absolute zero. Ratio data observations may thus be added, subtracted, multiplied, and divided. Height, weight, and accounting balances are such examples. Statistical tools discussed in later chapters work best for ratio data. Unfortunately, many types of data fall outside this classification.

Table 2.2 offers a visualization for the seven distinctions of data we've just covered. I offer this chart as warning that

Table 2.2 A Framework for Classifying Data and Scales

	Dichotomous	*Discrete*	*Continuous*
Nominal	Gender	Religious affiliation	
Ordinal	Pass/Fail grading	Olympic medals	
Interval		Day of week	Time of day
Ratio		Units sold	Price per unit

commonly taught tools of math and statistics work best on continuous, ratio data, where analysis makes full use of the number zero.

In your journey, you will come across data sets with variables and scales that span some or all of the categories in Table 2.2. For example, in studying the profitability of a retail store chain, you could receive a data set showing whether a given store location is owned or leased (dichotomous); the number of employees per store (discrete); location sizes in square feet (continuous); a rural, suburban, or urban setting (nominal); customer survey results on a not satisfied/satisfied/very satisfied scale (ordinal); average store temperature (interval) and profitability of a location (ratio).

Should you be forced to confront such complexity, then, as a numerate generalist, you should follow the basic methods described in subsequent chapters to identify *provisional* findings. You then must absolutely, positively bring in a propeller head to kick the tires and provide reassurance that any identified patterns are indeed supported by the data. Should you decide to invite several propeller heads to the party, be prepared for chaos. No two experts will likely agree on how to evaluate complex data sets.[1]

Let us now turn to an example that shows what we may do when the full power of zero is unleashed. In August 2015, the

[1]An illustration of the messiness of analyzing complex data sets is provided by a University of California at Los Angeles chart that maps statistical techniques to varied types of independent and dependent variables: https://stats.idre.ucla.edu/other/mult-pkg/whatstat/, retrieved 4 April 2021.

U.S. Securities and Exchange Commission (SEC), an administrative agency designed to protect the interests of U.S. investors, published rules requiring listed companies to disclose the median total annual compensation of their employees plus the pay of their CEO (U.S. SEC, 2015). This action responded to legislators' concerns that executive pay practices had contributed to the 2008 global financial crisis.

Table 2.3 presents selected disclosure from 2017 for a dozen firms regulated by the SEC. These data, making full use of zero, permit assessments of differences, measurements to high levels

Table 2.3 Compensation Rates for Select Companies in 2017

Company	*Median Employee Pay*	*CEO Pay*	*Ratio of CEO Pay to Median Pay*
Facebook	$240,430	$ 8,852,366	37
Netflix	183,304	24,377,499	133
Exxon	161,562	17,495,119	108
Goldman Sachs	135,165	21,995,266	163
Verizon	126,623	17,947,316	142
Ford	87,783	25,030,151	285
AT&T	78,437	28,720,720	366
JPMorgan Chase	77,799	28,320,175	364
GM	74,487	21,958,048	295
IBM	54,491	18,595,350	341
Coca-Cola	47,312	10,874,694	230
McDonald's	7,017	21,761,052	3,101
Representative value	$106,201	$20,493,980	193

Source: Data from Facebook, Google and Netflix pay a higher median salary than Exxon, Goldman Sachs or Verizon, By Rani Molla. Vox, Apr 30, 2018. Retrieved from: https://www.recode.net/2018/4/30/17301264/how-much-twitter-google-amazon-highest-paying-salary-tech, retrieved 15 Feb 2019.

of granularity, and calculations of ratios. As users of these data, we may make all sorts of comparisons.

From these continuous, ratio data, we may do many things. We may determine that the typical Ford worker made more than their counterpart at GM (87,783 – 74,487 = 13,298). We may perform analysis of CEO pay to seven significant digits (I have no idea why we would want to do so). We may also calculate and compare ratios (e.g., the CEO of Verizon made 142 times more than the typical worker there while the corresponding ratio at AT&T was 366).

As they say in direct-response TV ads, *But wait, there's more!* If we assume that this sample is representative of all large, American companies, then we may make inferences about U.S. corporate pay practices in 2017 for firms not in the data set. Assuming – and these are big assumptions – that we have reliable measures plus a representative sample, then we could argue that the typical employee of a large American company was paid about $100,000 per year in 2017, the typical CEO received about $20 million, and that the U.S labor market values the services of a CEO at 200 times that of the typical worker. We may also use this information to compare American corporate pay practices over time, across countries, or against forecasts. The possibilities are endless.

Before accepting that the use of zero meets everyone's needs, imagine that you are running errands while driving home from work. You pull over to drop off laundry at a dry cleaner. A parking sign warns that motorists must pay for weekday parking from 8 a.m. to 6 p.m. Your Internet-synched watch says it is 18:01, so you happily slide into an open parking spot and bypass the parking meter on your way to the store. As you return to your car, a member of the law enforcement community places a parking ticket under your windshield wiper.

You are blisteringly angry because you are a law-abiding citizen who has been falsely accused. To clear your good name, you must pay an undeserved fine or invest time to sort things out at

Table 2.4 Measures and Labels of Intelligence

IQ Range	*Midpoint*	*Label*
70–80	75	Borderline deficiency
50–69	60	Moron
20–49	35	Imbecile
0–19	10	Idiot

Source: Data IQ Basics, Retrieved from: iqcomparisonsite.com/IQBasics.aspx, on 3 Dec, 2018.

City Hall. You express outrage, but the officer does not bother to check his facts or listen to your grievance.

To vent your frustration, you need to share your belief that the officer is not very intelligent. To help you compose your thoughts, I offer you the reference chart provided in Table 2.4, which displays outdated labels for levels of intelligence, as measured by intelligence quotient (IQ) examinations and as described by labels associated with IQ ranges.

You have two choices. You could use continuous, ratio data to provide quantitative feedback that you believe the officer has an IQ of 50, half that of a normal person. Or, you could call the officer a moron, taking full advantage of emotive power of words. Personally, this is a no-brainer. I would call the officer a moron unless I was already having a bad day and resort to the word *idiot.*

This vignette shows the sterility of numbers. Conveying emotion, something we all need to do from time to time, requires the use of words. The Gettysburg Address's opening phrase *four score and seven years ago* invokes biblical imagery as Lincoln links America's heritage back to the founding fathers, much as the Hebrews traced their roots back to Adam and Eve in the Book of Genesis. Simple use of the number 87 would have fallen flat.

Martin Luther King Jr.'s brilliant use of metaphor in his "I have a dream" speech offers a calming vision of how the United States could become a color-blind society in a manner that could not have been achieved through reference to quantitative measures of racial inequality.

A numerate person remains a lifelong student of both numbers and words. Those able to draw on the power of both tools may identify relationships and then persuade others to take them seriously. Perhaps the sentiment was best expressed by a team of Swedish researchers who said that the world cannot be understood without numbers, and it cannot be understood with numbers alone (Rosling et al., 2018).

We close the chapter with a final example. I once heard a colleague say, *The numbers speak for themselves,* a statement I reject as patently false. Numbers, the raw material for numeracy, have no intrinsic meaning. Only by enforcing comparisons and then interpreting differences will one be able to extract information from data.

Suppose a firm closes its books and reports that it sold 2,000 units for the year ending December 31, 2018. This piece of discrete, ratio data makes admirable use of the number zero, yet it is completely useless. It is only through comparison and interpretation that meaning is revealed. Table 2.5 offers three of many possible comparisons.

On the left, we compare the number of units sold with that of the prior year. This year's sales represent a 25% increase, an impressive number given that the general economy grew at about 3% over the same period. The smiling face shows a favorable evaluation.

Table 2.5 Evaluation of a Number Depends on the Comparison

History			*Expectations*			*Competition*		
2018	*2017*	*Difference*	*Actual*	*Plan*	*Difference*	*Us*	*Them*	*Difference*
2,000	1,600	+25%	2,000	2,100	–5%	2,000	3,000	–33%

In the middle, we note that management had announced a year ago that it expected to sell 2,100 units in 2018. Actual results fell 5% short of the plan. The company failed to meet expectations set by senior executives. The flat face shows some level of concern that management was unable to deliver on its guidance. Perhaps this miss is a symptom of larger organizational problems.

On the right, we learn that a competitor was able to sell 3,000 comparable units in the same accounting period. Our performance falls a third short of that of a peer organization. The frowning face shows worry that our competitor may have a better business model, as reflected in, say, a superior product, distribution network, brand, or cost structure.

What's important is that the *same number* may give rise to happiness, caution, or discomfort, depending on the underlying comparison. The balance of this book reveals concepts that are based on the idea of comparing something with something else. Going forward, for the rest of your career, I ask that you never, ever share a number without enforcing a comparison.

Recap of Chapter 2 (Zero [0])

- 0 brings the ability to assess differences, measure with precision, and calculate ratios.
- Numbers facilitate comparisons but do not convey emotion.
- Numbers require comparisons to elicit meaning.

3

SAMPLE SIZE [n]

The second concept in our path to building a numeracy toolkit is represented by a lower case *n*, which signifies the ***sample size*** of a data set. This discrete, ratio measure counts the number of observations in a sample being studied and offers clues about the comfort we may take from conclusions reached from analysis of sample data. Perhaps every data set you will encounter in your career will be a sample of a larger, unknowable population. Whenever faced with a new sample, please reflect on its size, source, composition, and homogeneity before pushing buttons on a computer.

To begin, let's add three more words to our vocabulary.

Population is the complete data set for a variable of interest. As I write these words, an estimate for the number of people on our planet is about 7.6 billion. An exact count is unknowable given current measurement tools. Many people use a capital *N* to label population size. Almost every measure of *N*, regardless of domain, relies on estimates and is subject to uncertainty.

Sample, when used as a verb, means the process of extracting some observations from a population for further study. When used as a noun, sample means the set of observations that have been pulled from the population.

The number of observations in the sample is denoted with a lower case *n*, something that we are usually able to count with certainty. I note the alliteration of population parameter (e.g., *N*) and sample statistic (e.g., *n*) to remember how to distinguish these two summary measures of data sets.

We resort to samples because of time or budget constraints. I have yet to see a problem assigned where the boss says, *Relax. We have all the time in the world plus an unlimited research budget. Let's track down every single member of the population.* Kidding aside, it is almost impossible to know if a data set is truly complete. A defensible course of action is to assume that any data set we study is a sample of a larger population of unknown size.

Inference is the act of drawing conclusions about a population based on analysis of the sample. Inference is a form of inductive reasoning, where we reach generalizations from a study of specific things. Statistical inference puts forth predictions for phenomena that have not yet been observed.

Figure 3.1 ties these ideas together and introduces the issue of sample size. On the left is an imaginary opaque urn containing unobservable marbles. The population size (*N*) is 16, but we

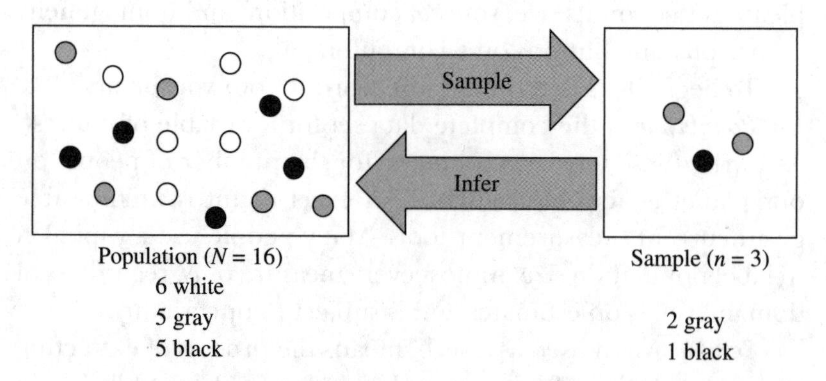

Figure 3.1 Good Inferences Require Sufficiently Sized Samples

don't know this. Our host shakes the urn to show that it contains a collection of marbles and then asks us to draw conclusions about the collection. To gather information, we are allowed to insert a hand into the urn to remove one marble at a time for visual examination.

We reach in, grab a marble, pull it out, and note that it is gray. We do this two more times, withdrawing a black marble and then another gray marble. The host stops us and asks us to make an inference about the urn's contents. Drawing on our growing vocabulary, we note that we have a small sample ($n = 3$) composed of discrete, nominal data. We see that the ratio of gray marbles to black marbles in our sample is 2:1 and infer that the urn contains a bunch of gray and black marbles, where the latter appear half as frequently.

Our host dumps the remaining marbles on a table. We realize that our conclusion is way off. Among other things, we missed the fact that there are white marbles that outnumber both the gray and black marbles. Further, the gray and black marbles exist in equal numbers. How embarrassing.

What went wrong was that our sample was too small. Good inferences rest on, among other things, a sufficiently sized sample. Well, you ask, what sample size is big enough? Without getting technical, I suggest a simple rule of thumb. If the sample size is below 30 ($n < 30$), then we have a "thin" sample. Thin means we're dealing with my so-called law of small numbers, which casts doubt on inferences reached from thin samples. In the marble example, when asked to offer a conclusion, I'd turn to the host and say, *Boss, we're working with three observations, a thin sample, so any inferences should be taken with a grain of salt. My early conclusion is that we have an urn filled with gray and black marbles in a ratio of two-to-one. However, I would like your permission to invest time and money to expand the sample. I'm hesitant to make any strong recommendations until the sample size reaches 30.*

In this case, expanding the sample would have allowed us to observe all the data points in the population, an unusual circumstance, and offer an answer with certainty. In the real world, sampling is costly in terms of money or time, so the boss may be hesitant to fund sample expansion. If we're told *No,* then the best we can do is to emphasize that inferences based on thin samples have a good chance of being wrong.

The parable of the blind men and the elephant reinforces the idea of thin samples. In one version, six blind men come across an unfamiliar elephant. Using only the sense of touch, each explores the unidentified object:

- The first man feels the tusk tip and presumes they have found a sharp arrowhead.
- The second touches an ear and decides there is a carpet.
- The third caresses the trunk and announces there is a snake.
- The fourth reaches a leg and reckons there is a tree.
- The fifth rubs a side and identifies a wall.
- The last grabs the tail and concludes there is a rope.

Each observation – with a sample size of 1 – leads to a different conclusion. No one offers an explanation that integrates the six data points. Efforts to convince others of the correctness of a particular interpretation bring two observers to the brink of a fistfight.

In the absence of a colleague with extraordinary critical thinking skills, the only solution is to expand the sample size. Additional observations permit further trial and error in efforts to make sense of raw data. The point is that small samples impair anyone's ability to reach sound conclusions. The presence of substantial variance (the topic of Chapter 5) among data points presses the need to expand sample size.

The opposite of a thin sample is a robust sample, but there is no bright line separating the two. If we have a sample with dozens of observations, then I start to be confident that we may begin to rely on inferences reached. However, I cannot say that the presence of an enormous sample puts me at ease. Really big data sets often aggregate observations that measure different things.

An example of working through sample size is asking a physician how many times they have performed a risky medical procedure. Responses of 1, 10, 100, and 1,000 elicit different reactions from a prospective patient. My great-uncle practiced obstetrics into his eighties. One could ask why young women would choose an old man to deliver babies. The answer is that he delivered 7,000 babies over a career and had experienced an enormous range of birthing complications. When n is in the thousands, we take some comfort in the quality of a doctor's analysis and conclusions.

Once we get a handle on the size of a sample, we need to consider its source. The dots we collect influence the dots we connect.[1] Our concern is ***sample bias***, where we gather a sample with a distribution of observations that differs from a corresponding distribution for the population. Sample bias brings bad inferences.

A famous example comes from the 1936 U.S. presidential election. Democratic President Franklin Roosevelt was running for reelection against Republican Governor Alf Landon of Kansas. *The Literary Digest,* a high-end general interest magazine, sampled American voters to predict the outcome of the November 3rd election.

The sample frame (source of the data) was three lists: the *Digest*'s own readers, registered automobile owners, and telephone users. The poll avoided the problem of a thin sample by collecting 2.3 million responses, an astronomically large n.

On October 31, 1936, the *Digest* published its survey results, which suggested that Landon would cruise to victory with 57% of the vote. Three days later, Roosevelt clobbered Landon, earning 62% of the vote. Table 3.1 summarizes results of the poll and election.

Despite having a huge sample, the *Digest* got it completely wrong. The problem was sample bias. Respondents to the poll,

---

[1]I lifted this line from William Lidwell's "Selection Bias" video module within the course *Universal Principles of Design* (LinkedIn Learning, 2015).

Table 3.1 Sample Size Is No Guarantee of Good Inferences

	Sample n = 2.3 million		*Population N = 44.4 million*	
	Responses	*Share*	*Votes*	*Share*
Landon	1,293,669	57%	16,679,543	38%
Roosevelt	972,897	43%	27,747,636	62%
	2,266,566	100%	44,427,179	100%

more likely to subscribe to a literary magazine, own a car, or have a telephone, did not reflect the nation's habits and living conditions in 1936, the middle of the Great Depression. Among the problems were selection bias (asking questions of people who are markedly different from the typical voter) and response bias (soliciting feedback from people more likely to voice an opinion).

Some introductory statistics textbooks implore readers to use random sampling to mitigate sample bias. One commonly suggested technique is to number each member of the population from 1 to *N*, determine the desired sample size, take a random number table, and then select observations with positions in the ordered data points that correspond with the random numbers.

This advice is ridiculous. First, we may not have any idea how many observations are in the population, so it's impossible to put them in order. Second, the population size may not be stable (for example, people may be born or die during the sampling process). Third, randomness is a construct that eludes a simple definition. Don't believe me? Strike up a conversation among propeller heads on how to collect a random sample from a given population and watch what happens.

Next, it may be difficult putting data points in an ordered list. Good luck, say, matching random numbers to fish captured from a lake. What do you do if one gets away? Further, you will typically want your sample to have the highest possible *n* and thus seek to use all available data, a topic covered in Chapter 12. Finally, you will often be forced to use data collected by others who do not have the skills or budget to create a random sample.

In doing your job, you may never work with random samples.[2] Instead, put your energy into assessing the degree of sample bias from the available data and then framing conclusions in terms of any identified biases. This framing process will necessarily be qualitative, but you will better serve your boss or client by being approximately right rather than precisely wrong.

One way to assess bias is to compare the frequency of another observed trait for both the sample and the population. So, let's say that we wish to estimate the average annual compensation rate of people aged 20–29 within a city. We collect both pay rates and gender for people in our sample. We then compare the proportion of male respondents with the fraction of population that is believed to be male, as given by a different data source. If the male proportion is 58% for our sample and 50% for the population, then we call out this difference when reporting results:

Boss, we performed a telephone survey of the annual pay rate of people in their twenties in our city. The sample size, n, was 100. The mean of all responses was $49,250. However, we should be cautious interpreting this result. First, 58% of our sample was male, but demographic data show that only about 50% of the people in this age group are male. Further, a phone survey requires that people are willing and able to answer the phone, so we may be at risk of response bias. Finally, we can't guarantee that we had the proper phone number for all people in our sample frame.

I believe this report meets the needs of our client. The survey method was fast, simple, and understandable. The report details the sample frame, sample size, and perceived types of bias involved.

---

[2]Interestingly, people asked to pick a random number between 1 and 10 often choose 7. See, for example, https://www.reddit.com/r/dataisbeautiful/comments/acow6y/asking_over_8500_students_to_pick_a_random_number/, retrieved 6 April 2021.

Openly describing our method and its weaknesses – all efforts at numeracy have weaknesses – does our boss a favor. The real value comes from the dialogue as the boss asks follow-up questions. They may want to know, for example, how we identified possible respondents, how many calls we made, what questions we asked, how many nonresponses there were, and so on.

The back-and-forth exchange of qualitative and quantitative information is what reduces uncertainty about how much people in our city get paid each year. Finding the "correct" answer would be prohibitively expensive, and this number would likely have changed by the time we compiled all the data from an exhaustive survey.

A truly numerate person considers what is *absent* in a sample's data set – no simple task. In Arthur Conan Doyle's short story "The Adventure of Silver Blaze," a racehorse disappears one evening. Sherlock Holmes solves the mystery after noting that a dog that was present that evening did not bark. The absence of expected data was evidence that the unidentified visitor was known to the dog.

My favorite example of coping with absent data came from World War II when statistician Abraham Wald was asked to study patterns of bomber damage inflicted by Axis anti-aircraft fire to propose adjustments to the positioning of armor plating to better protect Allied planes. The available data was the collection of planes that had survived battle and returned to base. The obvious solution was to add plating to areas associated with concentrations of bullet and shell holes.

Wald showed that this would be the wrong answer. The sample was composed only of planes that had made it back and thus distorted by survivorship bias. What he cared about was planes that were shot down and absent from the sample. Areas on planes with low incidence of holes indicated vulnerability because planes damaged there were less likely to return to base. The plane designers should add plating to areas in sampled airplanes with a comparative absence of damage.

Survivorship bias is prevalent when evaluating investment professionals. Investors want to place their money with money managers who have demonstrated an ability to earn superior returns. It is comparatively easy to evaluate money managers over a one-year horizon, but noise from capital markets' hiccups and blips may drown out evidence of superior performance over short time periods.

We could evaluate returns over a 10-year period, but we have a different problem. Only portfolio managers that stayed in business for more than a decade are included in the sample. Portfolio managers showing consistently weak results are more likely to suffer investment withdrawals and go out of business. Sample data for portfolio returns extending over 10 or more years are almost certainly overstated due to survivorship bias.

After considering sample size and bias, we need to turn our attention to possible interdependence among things in a sample. When you first studied statistics, your instructor likely used inanimate objects like coins, cards, and dice to show how samples give rise to inferences.

The benefit of using lifeless objects is that, in the absence of weird quantum physics, one item does not affect another. The result of a coin flip is not influenced by the result of the previous trial. As they say in Vegas, the roulette wheel has neither a memory nor a conscience. When the behavior of one thing in a sample has no influence on the behavior of another, then we have independent trials – a necessary condition for using the normal distribution, the topic of Chapter 6.

When the sample under consideration involves people or measures of the behavior of people, then we need to be sensitive to the possibility of interdependence when drawing statistical conclusions. People watch each other, talk to each other, learn from each other, and thus influence each other (Andriani & McKelvey, 2007). Human behavior within groups defies simple explanation due to feedback loops. In such cases, we need to be cautious when making inferences from samples drawn from past behavior.

A classic example of failing to consider interdependence was the Long-Term Capital Management (LTCM) hedge fund fiasco. Some brilliant traders and economists (including two Nobel laureates) used samples of prices from related securities to develop trading rules in the face of foreseeable market fluctuations. To boost returns, the fund borrowed money to juice payoffs from exploiting tiny differences in related security prices.

The group's mistake was failing to consider that security prices result from the actions of human beings who watch each other, talk to each other, and so on. In August 1998, the Russian government unexpectedly defaulted on its debt. Investor panic brought unforeseen changes in market prices that were not contemplated by the developers of LTCM's trading rules. Because the fund had borrowed so much money, the money managers could not ride out the storm. LTCM collapsed and was subsequently liquidated.

In my opinion, the geniuses forgot that human behavior is interdependent and thus sometimes unstable. It can be dangerous to make big inferences on samples measuring fickle human actions.

When you look at a data set, consider what it is we're measuring. If we're studying what appear to be independent, inanimate phenomenon, then we're probably safe making inferences from samples that are sufficiently large and unbiased. However, if we're measuring things that are part of a system with feedback loops, then we should temper conclusions. We could say, *Boss, we've studied blue jean purchasing by millennials. The team did its best to get a sample that is decently sized (n = 200 people) and mitigates sample bias (the distributions of age, location, and socioeconomic status for buyers in the sample correspond with outside demographic data). However, let's keep in mind that we're studying people with interdependent behaviors. New fashion trends can appear out of nowhere and spread like wildfire. Something not contemplated in our analysis could trigger an abrupt change in consumer tastes.*

A related problem of interdependence is the observer effect, where the act of observation disturbs the subject. Using a tire gauge simultaneously measures air pressure and changes its level. Observer effects complicate efforts to measure things (King, 2018).

These cautionary words force the client to consider factors not considered in the statistical study. Showing humility by emphasizing unavoidable qualitative issues is the best way the team may serve its customer. As noted earlier, we understand the world neither without numbers, nor with numbers alone.

Finally, let's consider heterogeneity, the fourth issue with samples. Statistical analysis aggregates individual observations to form inferences. The best way to avoid the law of small numbers is to increase n. If we gather a ton of data, we may argue that our conclusions should be taken seriously because we have a robust sample. Information provided by our analysis reduces uncertainty and allows us to deliver conclusions with greater confidence. Not so fast.

Our analysis needs to consider the sameness of the data points in our sample. Have we really collected a group of like apples? Or have we inadvertently allowed some figurative oranges into our study? If we were to gather data on new car prices over multiple decades, we have an issue because both units of measure have changed. Compared to 30 years ago, cars have more technology and safety features today while the purchasing power of a dollar differs due to changes in prices, quantities, and qualities of goods and services sold over the past three decades.

Consider the problem of studying Olympic gold medal winners for the 100-meter dash. The race has not changed since 1896, when the modern Olympic era began. However, consider the data in Table 3.2.

At first glance, it appears that we have 28 interchangeable data points: we're measuring male athletes running identical distances. Yet, further examination reveals that each observation has its quirks.

Table 3.2 Winners of Olympic Men's 100-Meter Dash

Year	*Winner*	*Location*	*Seconds*
1896	Tom Burke	Athens	12.00
1900	Frank Jarvis	Paris	11.00
1904	Archie Hahn	St. Louis	11.00
1908	Reggie Walker	London	10.80
1912	Ralph Craig	Stockholm	10.80
1920	Charles Paddock	Antwerp	10.80
1924	Harold Abrahams	Paris	10.60
1928	Percy Williams	Amsterdam	10.80
1932	Eddie Tolan	Los Angeles	10.38
1936	Jesse Owens	Berlin	10.30
1948	Harrison Dillard	London	10.30
1952	Lindy Remigino	Helsinki	10.40
1956	Bobby Morrow	Melbourne	10.50
1960	Armin Hary	Rome	10.20
1964	Bob Hayes	Tokyo	10.00
1968	Jim Hines	Mexico City	9.95
1972	Valeriy Borzov	Munich	10.14
1976	Hasely Crawford	Montreal	10.06
1980	Allan Wells	Moscow	10.25
1984	Carl Lewis	Los Angeles	9.99
1988	Carl Lewis	Seoul	9.92
1992	Linford Christie	Barcelona	9.96
1996	Donovan Bailey	Atlanta	9.84
2000	Maurice Greene	Sydney	9.87
2004	Justin Gatlin	Athens	9.85
2008	Usain Bolt	Beijing	9.69
2012	Usain Bolt	London	9.63
2016	Usain Bolt	Rio de Janiero	9.81

The races run in 1896, 1900, and 1904 were measured to the nearest whole second; the next five, to tenths of a second; and the balance, to the nearest hundredth. Obviously, the technology used to measure time elapsed changed over a century. Note the big drop in 1968. The race that year was held in Mexico City, where the stadium had an elevation of 2,240 meters (7,350 feet) above sea level. Runners in that venue faced lower air resistance. The winning time in 1980 was slower than that for the five preceding contests, perhaps evidence of diminished competition as certain countries boycotted the games in Moscow to protest the 1979 Soviet invasion of Afghanistan. Other variables such as changing shoe technology, track surfaces, starting blocks, and training regimens complicate efforts to draw conclusions. Students of this Olympic event point out why each race has its own particular circumstances. Rarely do data sets have fungible observations.

My favorite warning for the perils of data aggregation come from an analysis of health care research written by a medical journalist. She attended many conferences and had trouble sorting through conflicting findings discussed by physicians.

> And then one day it hit me: They are looking at different universes. Time after time, investigators claim to be researching a specific "disease," but each investigator defines this disease in a slightly different way. Without a standard definition, no study may be compared with any other. (Ziporyn, 1992, p. 2)

Differences among observations within a sample confound the ability to draw conclusions from statistical analysis. I have seen this problem many times in my own field of accounting, where different companies calculate operating earnings in as many ways (King, 2017).

Recap of Chapter 3 (Sample Size [n])

- Analysis of samples give rise to inferences about larger populations.
- Any inference made from a sample should include qualitative evaluation of:
  a. Sample size;
  b. Sample bias, where the sample is not representative of the population;
  c. Interdependence of observations within the sample; and
  d. Heterogeneity among observations in the sample.

4

SAMPLE MEAN [x]

This chapter introduces averages, the third milestone in our path to numeracy. The lower case letter x signifies the arithmetic mean, the doyenne among many measures of average, for observations within a sample. We would use the Greek letter μ if we meant the corresponding figure for an entire population. We often use x to develop an estimate of μ, something we discuss in Chapter 10.

We need averages to make comparisons within and across groups. An ***average*** collapses a collection of numbers into a simple summary measure. I could show you a mind-numbingly long list of employee names and annual pay rates for a large company, or I could present you with a sheet of paper with the notation $n = 43{,}255$, $x = \$51{,}003$.

My brief report conveys that my data set contains 43,255 observations. I did not characterize the data set as a population (by using an upper case N) because I could not guarantee that I had (1) tracked down every person who worked at the firm when the measurement was taken or (2) sorted through thorny classification issues of who was an employee, contractor, or temporary worker as of the measurement date. The sample size is well above our cutoff value of 30 and thus not thin. The report would be improved by providing commentary on how I collected the data and defined the term *employee.*

The report also says that the arithmetic mean is $51,003 per year, the pay rate for an imaginary "average" employee. I could then use this measure to enforce a comparison with another employee, the average for all employees from the same company last year, for the average of employees from a competitor firm, against a previous forecast, and so on. The primary idea is that averages facilitate comparisons within and across groups as part of efforts to make predictions or provide explanations.

There are many kinds of averages used to summarize data sets. In this chapter we focus on five. Each measure has shortcomings, so a numerate person serves their client by highlighting problems with any chosen measure of central tendency. The use of averages is a necessary, incomplete step to extract information from data.

Let's introduce three commonly used types of averages in a contrived example provided in Figure 4.1. Imagine that a primary school teacher asks her class of 26 students to read books over the summer and report the number of books completed. The teacher takes the book counts, sorts them by student name and by number of books read, and considers how she may assess (1) the diligence of this particular class relative to those of prior years and (2) which students within this year's class were particularly industrious.

The teacher notes that she has a thin sample ($n = 26$), so she's hesitant to draw sweeping inferences from this collection of observations. She also sees that she has ratio data – there is an absolute zero in this scale – and is thus able to perform divisions when evaluating the data. It is certainly possible for students to have read fractions of a book, but no student reported doing so. Thus, we seem to be working with discrete data. As with all data sets, we wonder what is not recorded in the sample.

	Sorted Alphabetically		Sorted Numerically		
Observation Number	*Student*	*Number of Books Read*	*Student*	*Number of Books Read*	
1	Amanda	6	George	0	
2	Beth	13	Dan	2	
3	Charlie	17	Ulysses	3	
4	Dan	2	Rachel	4	
5	Emily	13	Frank	5	
6	Frank	5	Paul	5	
7	George	0	Sam	5	Mode = 5
8	Harry	7	Tony	5	
9	Iris	15	Zack	5	
10	Jane	7	Amanda	6	
11	Kim	8	Quinn	6	
12	Lester	21	Harry	7	
13	Marisa	17	Jane	7	
					Median = 7.5
14	Nancy	10	Wally	8	
15	Oscar	11	Xander	8	
16	Paul	5	Kim	8	
					Mean = 9
17	Quinn	6	Nancy	10	
18	Rachel	4	Oscar	11	
19	Sam	5	Beth	13	
20	Tony	5	Emily	13	
21	Ulysses	3	Iris	15	
22	Violet	17	Yvonne	16	
23	Wally	8	Charlie	17	
24	Xander	8	Marisa	17	
25	Yvonne	16	Violet	17	
26	Zack	5	Lester	21	
		234		234	

$$\text{Mean} = \frac{\text{Sum of all books read}}{\text{Number of students}} = \frac{234}{26} = 9$$

Figure 4.1 Mean, Median, and Mode Are Three Measures of Central Tendency

The teacher starts with the ***arithmetic mean***, calculated by dividing the sum of observations by the number of observations ($x = 234/26 = 9$). The mean, to use an abbreviated label, provides a value for an imaginary, typical student who read nine books last summer. Excel uses the command =AVERAGE to calculate the mean. Please do not use the words "mean" and "average" interchangeably: the arithmetic mean is one component of a set of statistical tools used to measure averages.

One way to think of the mean is to envision a thin, stiff board with evenly spaced marker points labeled from 0 to 21. We place a single coin on 0 (representing the book count for George), an identical coin on 2 (Dan's count), one coin for 3 (Ulysses), one coin for 4 (Rachel), five coins on 5 (one for Frank, Paul, Sam, Tony, and Zack), and so on until we place a single coin on 21 for Lester, who had been especially productive. If we were to place this arrangement on a fulcrum, the board would balance when the point is placed under marker number 9.

Note that x is an idea, not a tangible thing. In this case, no student in the class read nine books. We simply use x as a reference point should we wish to compare this class's performance with another. We should be cautious when using x to make a comparison with a specific student because the average person does not exist (Rose, 2016). For example, saying that the average person has one testicle and one ovary is just as true as it is useless.

The teacher then calculates the ***median*** number of books by selecting the midpoint in the list sorted by number of books read. In our sample, there was an even number of observations, so she calculates the midpoint from the arithmetic mean of observations 13 and 14 (median $= [7 + 8]/2 = 7.5$).

The median signifies the dividing point that separates observations into halves, with one above and the other below this measure of central tendency. Our median value of 7.5 books creates a 50/50 split of data points. As with mean, the median is an idea instead of something tangible. In our data set, no student read 7.5 books. Excel uses the command =MEDIAN to calculate the median of a data range.

A key distinction between mean and median is that they react differently to the presence of ***outliers***, unusual observations far removed from the mean, something we study in Chapter 7. Suppose that productive Lester had read 47 books instead of 21 last summer; then the class's recalculated mean rises to 10, while the median remains unchanged at 7.5.

Means are less stable than medians when a sample contains an outlier with a very large or small observation. As an example, consider the problem of measuring the mean wealth of a group of donors at a fundraiser just before and after Bill Gates walks into the room.

Our third measure of central tendency is the ***mode***, which identifies the most commonly appearing value of observations in the data set. In this case, the mode is 5, as evidenced by five different students completing this number of books. The analogy for mode is a vote for an election. The winning candidate is typically the one who receives the greatest number of votes.

Mode is a less useful measure of average because a sample always has one mean and one median but may have no mode, a single mode, or multiple modes. However, mode is the only measure of central tendency available when one is confronted with nominal data that may not be sorted by value. Examples of nominal data include hair color, political party affiliation, and blood type.

Excel uses the command =MODE.SNGL to identify the mode in a column of data with numeric values, but this can be tricky. Some data sets have an absence of a mode (so the software will return an error notification) while other data sets are multi-modal. For example, this collection of observations has three modes:

0, 1, 1, 2, 3, 3, 4, 5, 5, 6

Unless one is careful, the software (or the versions I've used) will return the smallest numeric value within the modes (in this case, 1). An Internet search will provide methods to cope with this problem. The point to remember is that a sample may have no, one, or multiple modes.

In Figure 4.1, we have three different measures of central tendency. None of them is the "right" answer to the question of the average number of books read last summer. The use and interpretation of measures of average require judgment.

A fourth type of average is ***geometric mean***, which is used to calculate average rates of change for variables that are multiplied together. In the previous example, books are objects that may be added together to evaluate reading performance across students. By contrast, investment results use rates of return that must be multiplied against principal invested to evaluate investment performance over accounting periods.

To make this clearer, consider what would go wrong if we tried to calculate the arithmetic mean of annual returns for an investment across two years. Assume we invest $100 in a security that initially stumbles and loses 10% in the first year and then rallies to provide a 30% gain in the second year. If we start with $100, our investment drops to $90 after year 1 ([1 – 10%] × $100 = $90) and then grows to $117 after year 2 ([1 + 30%] × $90 = $117).

Taking the arithmetic mean of these two returns ([–10% + 30%]/2) provides an answer of +10% per year, which is wrong. If the security had delivered a 10% average annual over two years, then the investment's ending value would have been $121 ($100 × 1.10 × 1.10). The appropriate average in this case, the geometric mean, is about 8.2% per year ($100 × 1.082 × 1.082 = $117). We need to use the geometric mean because the amount of money invested varied from one year to the next.

Figure 4.2 shows how to use Excel to calculate this figure. Column E shows the annual investment return for a change in a security's price (percentage change = [Price_t / Price_{t-1}] – 1), and column G adds the investment return to 1. Cell G7 shows the calculation of geometric mean over an investment horizon using the command =GEOMEAN(data range) – 1. We subtract 1 from the geometric mean to present only the average return.

G7 | =GEOMEAN(G4:G5)-1

	A	B	C	D	E	F	G
1							
2	**Year**		**Price**		**Return (i)**		**1 + i**
3	0		$ 100		n/a		n/a
4	1		$ 90		–10%		0.90
5	2		$ 117		30%		1.30
6							
7					Geometric mean		8.2%

Figure 4.2 Use of Excel to Calculate Geometric Mean
Source: Used with permission from Microsoft Corporation.

Geometric return calculations require evenly spaced time intervals across observations.

The fifth type of average discussed in this book is ***harmonic mean***, the average rate for a sequence of steps that have identical effort. Let's say you wish to measure the average mileage for a car that travels up a steep hill for 100 miles, burning gasoline at the rate of 10 miles per gallon, and then travels downhill for the next 100 miles, sipping fuel at a rate of 100 miles per gallon.

Your intuition may suggest that the average mileage appears somewhere in the middle between the two observations of 10 and 100 miles per gallon. Your intuition would be wrong. The answer is the harmonic mean of 18.2 miles per gallon, calculated by dividing the 200 miles traveled by 11 gallons of gasoline consumed.

You came across this type of problem in high school when you were asked to answer questions such as, "If Bill is able to paint a house in two days while Ted is able to do so in three days, how would it take for the two of them working together to get the job done?" Assuming they don't interfere with each other, the answer is 6/5ths of a day because, in one day, Bill completes 1/2 of a house while Ted completes 1/3. Adding these two rates together gives us a combined rate of 5/6 of a house in day.

Dividing the job length of one house by the combined rate of 5/6 of a house per day gives an elapsed time of 1.20 days.

Analysts use the harmonic mean to study sequential tasks of equal lengths. For example, one would use this measure to compare average speeds of runners within relay teams competing in the 4 × 100-meter dash. Government regulators set U.S. annual passenger car fuel economy goals using the harmonic mean.

The Excel command for this measure is =HARMEAN. Figure 4.3 shows how Excel may be used to calculate the harmonic mean for average mileage of a car traveling up and then down a 100-mile hill. Calculation of harmonic means requires equal units of work or activity for each observation.

At this point, we have introduced five of many ways to calculate an average. Space limitations and your waning interest cause me to stop. The key idea is that there is not a single, correct way to measure averages. Table 4.1 presents a small sample of observations together with the average using the five methods we have introduced. Notice that each tool provides a different measure.

Judgment for which (if any) of these measures should be used to calculate a group's central tendency is part of the craft of numeracy. I cannot offer universal advice other than urging you to never refer to a given measure of central tendency as "the average."

G6 | =HARMEAN(G3:G4)

	A	B	C	D	E	F	G
1							
2	**Direction**		**# Miles**		**Gallons**		**Miles/Gallon**
3	Uphill		100		10		10
4	Downhill		100		1		100
5							
6					Harmonic mean		18.2

Figure 4.3 Use of Excel to Calculate Harmonic Mean
Source: Used with permission from Microsoft Corporation.

Table 4.1 Many Ways of Calculating an Average

Observation	*Value*
1	10
2	20
3	30
4	40
5	40
6	50
Arithmetic mean *(x)*	31.7
Median	35.0
Mode	40.0
Geometric mean	28.0
Harmonic mean	23.7

Let's go back to Lester, who read 21 books and seems to be a bit unusual. Is the distance from the number of books he completed from the mean (21 – 9 = 12) so great that he should be characterized as an oddball? Answering this question involves comparing the distance of extreme values with the mean and then determining whether the distance is so great that the observation should be classified as an outlier.

Consider what happens if Lester had gone crazy and read two books a day over his three-month summer vacation. If he had read 182 books, the mean for the entire class would have risen from 9.0 to 15.2 books, a huge jump. In this case, swapping out one large value for an even larger one boosts the mean without having any influence on the median (half of the class still read fewer than 7.5 books) or the mode (5 remains the most observed value).

If we are interested in extreme performance, we should pay more attention to means because this measure is the most sensitive to the presence of outliers. If we are more interested in the performance of representative members of a group, the use of medians is more helpful.

If we believe that 182 represents measurement error (Lester lied about the number of books he read or gave himself credit for skimming certain books), then we should drop his observation from the sample and recalculate the desired measures of central tendency for the remaining 25 observations.

Concern for the middles or the extremes depends on the audience. Scientists tend to focus on central tendency while writers of business books tend to focus on extreme performance (Andriani & McKelvey, 2007). After all, who wants to read business books about average companies? If we're faced with nominal data, we have no choice and must resort to the mode. If we're exploring data sets with varied growth rates over time, then the geometric mean should probably be emphasized. If what we study involves sequential tasks of equal length, then the harmonic mean is perhaps the best measure.

When evaluating survey data ("On a 1-to-5 scale, how would you rate this car dealer?"), the median is probably the best measure of central tendency because such scales do not provide interval data associated with the arithmetic mean. Let's be honest, how often have you candidly given the lowest possible customer service rating when you were displeased? I don't, because the last thing I want is to answer follow-up questions from a management team seeking to understand my unhappiness. In this case, there is a bigger difference from dropping from 2 to 1 than there is from 3 to 2.

Table 4.2 summarizes my views of when to use the five averages discussed in this chapter.

Sometimes, we will work with a large sample of continuous data recorded to a sizeable number of significant digits. An analyst may want to know if clumps of observations gravitate near certain values or if observations are spread evenly over the sample's domain. An attempt to find modes proves to be ineffective because few, if any, observations share the exact same value

Table 4.2 Summary Descriptions for Five Types of Averages

Average	*Description*	*Typical Use*
Mean	Balancing point	Consider influence of outliers
Median	50/50 split	Ignore influence of outliers
Mode	Most common	Examine nominal variables
Geometric mean	Growth rate	Compute investment returns
Harmonic mean	Interval rate	Evaluate sequential tasks

in this robust sample. For example, data sets listing historical foreign currency exchange rates are often calculated to four or more decimal places and may not have any repeating values.

In these situations, one should create a ***histogram*** to visualize a frequency distribution, a graph where the horizontal *x*-axis shows different values of the variable of interest while the vertical *y*-axis shows counts for how frequently a particular range of values occurs.

Histograms accomplish this task by creating ***bins***, buckets that collapse continuous data into discrete data ranges. An example is classifying this sequence of grades achieved by 15 students in a class:

36 45 51 56 64 69 73 74 83 86 89 92 94 97 99

No two students earned the same grade, so there is no mode. However, we may aggregate student scores into bins labeled Fail (bold scores up to 69) and Pass (remaining scores of 70 and higher) to determine that 60% of the class earned a passing grade. A histogram would show this relationship in a bar chart displaying the discrete categories of Fail and Pass on the *x*-axis and counts of the students within each bin on the *y*-axis.

Figure 4.4 shows a histogram for continuous data that have more than one relative mode. The data come from 750 completed games scored at a bowling alley on one day. It would be

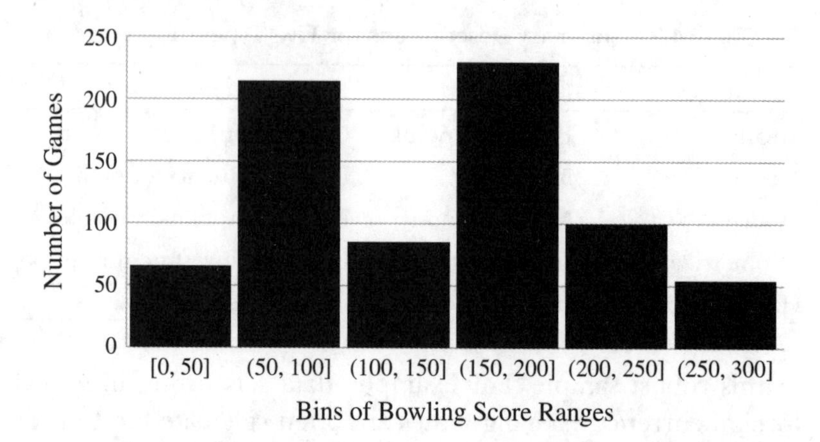

Figure 4.4 Histogram Showing Distribution of Bowling Scores in One Day

easy to calculate the mean, median, and mode for this data set, but such effort would not reveal the full story.

The histogram aggregates individual scores into six bins, each with a width of 50 points (the worst score possible in a game is 0; the best, 300). The *y*-axis in this frequency distribution counts the number of games associated with each bin.

The picture suggests that there were two types of bowlers who came to the bowling alley that day. One group appears to be casual bowlers who have a tendency to earn modest scores below 100 while a second group appears to be more experienced (or serious) bowlers who tend to earn scores that are about 100 points higher.

When using bins, brackets – [ ] – signify that a number in the range should be included in the bin, while parentheses – ( ) – show that the bin includes all observations that approach but do not include that number. Thus, the bin notation (50, 100] states that this bucket should include all games from a score of 51 through 100 points. Games with 50 points are placed in the preceding bin.

We will return to histograms when we study the normal distribution in Chapter 6.

We close out this chapter by elevating awareness of how measures of average change as we modify the frame of reference for a data set. Averages are context-specific ideas, not immutable laws of nature. An example of how context shapes findings is ***Simpson's Paradox***, a situation where a trend disappears or reverses when groups are combined.

A claim of gender bias was lodged against the University of California, Berkeley, in the 1970s. Aggregate admissions data showed that the proportion of women applicants admitted to graduate programs was lower than the corresponding percentage for men. An implication is that admissions boards discriminated against women applicants.

The resulting investigation revealed that the overall claim could be substantiated; however, examination of more program-specific data revealed that women tended to apply to more competitive programs with lower acceptance rates (Bickel et al., 1975). Within these more competitive programs, women earned higher rates of acceptance than men. The gender difference reversed upon examination of more granular data.

A more recent example comes from trying to answer the question of whether Derek Jeter or David Justice, two baseball greats, was the better batter in the mid-1990s. A measure of success is whether a batter achieves a hit for an attempt, defined as an "at bat." Batting averages are like bowling scores, where higher values represent more favorable outcomes. Table 4.3 shows comparative statistics for the 1995 and 1996 baseball seasons.

David Justice earned a higher batting average in each of the two seasons. However, Derek Jeter posted the superior average for the two-season period. This reversal stems from the weighting of the two averages in the combined statistic. Jeter's worse year and Justice's better year each had a comparatively low number of

Table 4.3 Simpson's Paradox: Conclusion Changes as Data Are Aggregated

	1995			*1996*			*Combined*		
	At bats	*Hits*	*Average*	*At bats*	*Hits*	*Average*	*At bats*	*Hits*	*Average*
Derek Jeter	12	48	0.250	183	582	0.314	195	630	0.310
David Justice	104	411	0.253	45	140	0.321	149	551	0.270

Source: Based on "Simpson's paradox." Wikipedia, 20 Feb. 2019. Web. 2 Mar. 2019.

at bats and hits. Baseball fans in my orbit give the nod to Jeter as the better batter.

Calculating averages is a necessary first step to making comparisons within and across data sets. This chapter shows that there is no single, correct way to do so; it also emphasizes that averages are constructs that do not reflect specific data points.

Recap of Chapter 4 (Sample Mean [x])

- Averages are abstractions that facilitate comparisons within and across groups.
- Different types of averages may yield different measures of central tendency.
- Histograms convert continuous data into discrete bins to assess relative frequencies.
- Averages may change as the analyst modifies what goes into a sample.

5

SAMPLE STANDARD DEVIATION [s]

In the previous chapter, we noted how an average condenses observations into a single, representative number. Any such winnowing brings a loss of information. Perhaps the most important loss is any sense for how far individual data points stray from a measure of central tendency.

We've all heard the parable of the man who drowned in a river with an average depth of three feet because a deep trench in the middle offset shallow banks. This is an example of the problem of ***variance,*** where a data set contains observations that are far removed from the average, diminishing its usefulness. A high level of variance makes an average downright dangerous if we seek to use it to predict or explain something.

Relying on averages buffeted by variance could ruin your day or stunt your career. A decision that contemplates central tendency alone (say, staffing for a mean of 10 customer service complaint calls received per hour) could bring heartache should extreme observations rear their ugly heads (receiving over 100 calls per hour for an extended period).

Statisticians have many tools to assess variance. Perhaps the most accepted way to quantify variance is ***standard deviation***, a measure showing the typical distance of an observation from a data set's arithmetic mean. We use the lower case letter s when discussing standard deviation of a sample and the Greek letter σ (sigma) when discussing standard deviation of a population. Numeracy studies incomplete samples, so we emphasize s over σ for the remainder of the book.

Suppose your spouse, who is entertaining a buddy, whispers to you that it would be nice if the three of you could each have an apple as a snack. This may be a bad idea because you're not sure how many apples are in the refrigerator. The household's shopping and eating cycles have irregular intervals, so estimating apple counts at a point in time is no easy task.

One way to reduce uncertainty is to reflect on recent trips to the fruit drawer. You draw upon your superpower of combining amazing recall with mental math to note that the last six times you opened the drawer you counted, respectively, 1, 2, 3, 4, 6, and 8 apples. Resulting sample statistics are $n = 6$ and $x = 4$.

We could use $x = 4$ as a prediction for the unknown number of apples on hand now. If we're right, we would have enough apples on hand to come across as good hosts; if we fail to have three apples on hand, we risk embarrassment. Sample standard deviation offers help in sorting through this uncertainty.

Table 5.1 shows how to calculate s using n and x. Working our way from left to right, we first list all observations and the corresponding apple counts (column A). We then calculate the distance each observation falls from the sample mean of 4 (the value of column A less the value of column B equals the value of A – B). So, observation 1 with a single apple brings a difference of –3 apples (remember, numeracy means working with ideas, not tangible things).

We're interested in the total of all differences from the mean. However, it does not make sense to add up entries in the Differences from Mean column because positive and negative values

Table 5.1 Calculating a Sample Standard Deviation

		(A)	*(B)*	*(A) – (B)*	*[(A) – (B)]²*
	Observation	*# Apples*	*Mean of Data [x]*	*Difference from Mean*	*Squared Differences from Mean*
	1	1	4	–3	9
	2	2	4	–2	4
	3	3	4	–1	1
	4	4	4	0	0
	5	6	4	2	4
# observations [n]	6	8	4	4	16
		Sum of squared differences from mean			34

$$\frac{\text{Sum of squared differences from mean}}{(n-1)} = \frac{34}{5} = 6.8 = \text{Sample variance}$$

$$\text{Square root of sample variance} = \sqrt{6.8} = 2.6 = \text{Standard deviaion}[s]$$

will offset. In this case, the sum of differences equals 0, which grossly understates any attempt to assess variance.

The trick is to square the differences, which we do in the last column on the right. For example, the squared difference for the first observation is $-3 \times -3 = 9$. In this fashion we convert all differences, both positive and negative, into positive values, which share the same sign. We may now use addition to accumulate all observations' distances from the mean. In our case, the sum of squared differences from the mean is 34.

The next step is to calculate the arithmetic mean of the squared differences to develop a summary measure of differences. We do not take the easy step of using the sample size ($n = 6$) as the denominator to calculate the arithmetic mean. Instead, we make an adjustment that reduces n by 1 ($n - 1 = 5$), lowers the size of the denominator, and thus raises the value of the resulting average to reflect uncertainty from working with a sample. If we had a population – and the comfort from working with a complete data set – then we would divide the sum by N.

In our example with a sample of observations, we divide 34 by 5 to arrive at an average of 6.8 apples^2 ("apples squared" – again, we're working with ideas, not tangible things). Statisticians use the term *variance* to label this ratio. My informal, practitioner background causes me to use the word *variance* casually to describe the phenomenon of observations straying from their arithmetic mean. My conversations with propeller heads have run amok when I use *variance* in an informal manner.

The unit of measure of apple^2 is difficult to visualize. To cope, we take the square root of this ratio to arrive at a standard deviation of 2.6 apples. The square root undoes the squaring we performed in the far-right column.

In Excel, we can dispense with this drudgery and simply use the command =STDEV.S to calculate the sample standard deviation for a group of observations. As an aside, a standard deviation must always be positive (we take the square root of squared values), so if you ever see a negative sign associated with a standard deviation, you know that someone made an error.

Armed with the three core summary statistics (n = 6, x = 4, s = 2.6), we're now able to help your spouse. We note that the typical observation strays 2.6 apples from the mean. Let's just round to the nearest whole apple and say that the average distance is three apples. We could create a confidence interval from 1 to 7 apples ($x \pm s = 4 \pm 3 = [1, 7]$), which shows that there is a decent chance that we have fewer than three apples in the drawer.

You could say to your spouse: *Honey, offering apples as a snack is not a good idea. Drawing on six observations, I note that the mean number of apples on hand is four and the standard deviation is three apples. The thin sample suggests that we shouldn't take much comfort from the average as a predictor of apple count, and the substantial level of variance relative to the mean brings additional concern whether we may use* X = *4 to make a prediction. I have sizable doubt that we have at least three apples in the fridge right now. Let me make some popcorn.*

This contrived conversation would never happen, but it shows the reasoning involved in numeracy. We start with incomplete data, calculate some basic statistical measures, and then use them to develop a practical conclusion that reduces uncertainty. The conclusion, of course, could be wrong. As we said earlier, statistics means never having to say you're certain.

The important point is that we have arrived at the three foundational sample statistics: n, x, and s. Developing the craft of numeracy requires gaining comfort analyzing relationships among these three measures.

You already have an intuitive sense for how to do this. Suppose that an online rental home displays 35 reviews, where every review has the maximum score of five stars. Assuming that the reviews are valid, you would take a lot of comfort from these data.

The average review must be five stars (or close to it), and there must be little variance among reviews because weaker ratings would depress the average below five. You understand but may not formally recognize that n = 35, x = 5, and s = 0. We have a robust sample with a high score and little variance. It is probably safe to use the five-star rating to enforce comparisons against other properties and to make a prediction about our vacation experience should we rent the house. *Honey, we should rent this home for our vacation.* We did and had a great experience.

By the time that you finish this book, I hope that you will be amazed by the scope of reasoning afforded by analysis of these three simple numbers. When handed a collection of observations comprised of ratio data, please calculate these three sample statistics and then think about them individually and in the aggregate:

- **n** Sample size assesses the degree of comfort we may take from our analysis of the data. A thin sample ($n < 30$) exposes us to the risk that the sample is not representative of an unobservable population. Figure 3.1 reminds us that good math and impeccable sampling may bring inaccurate conclusions if we're limited to a handful of observations.
- **x** The arithmetic mean offers a summary measure of central tendency based on the balancing point of all observations. The use of x offers a quick way to enforce a comparison within or across groups or to make a prediction for a yet-to-be observed observation. The caution of working with x is that its value moves with the presence of outliers. A smaller n suggests that x may have been substantially influenced by one or two extreme values.
- **s** The sample standard deviation shows the typical distance a given observation strays from x. Increasing levels of variance degrade the usefulness of averages. A high standard deviation may make it dangerous to rely on x when enforcing comparisons or making predictions. We should worry more about the relationship of s to x when we have thin samples, a topic explored in Chapter 10.

In class, I go through the exercise of putting three values on the board (for example, $n = 120$, $x = 10$, $s = 3$), and then have students offer interpretations. I then swap out values (say, reducing n to 12 or increasing s to 30) to see how the interpretation changes.

With a little practice, the students display more sophisticated statistical reasoning than what I heard from most of the highly paid investment bankers who called on me during my time in industry.

This type of analysis requires that we add another term to our growing vocabulary. ***Coefficient of variation***, the ratio of standard deviation divided by arithmetic mean, assesses the magnitude of variance relative to the average observation. So if $x = 4$ apples and $s = 2.6$ apples, then s/x equals 65%.

The ratio s/x provides a pure number because the units of measure cancel out (apples divided by apples gives an absence of units). A unitless measure allows us to compare variance across samples with different units of measure, time periods, and sample sizes. All you need to know is that the higher the ratio, the more dangerous the average.

Table 5.2 offers my nonscientific scale for evaluating the coefficient of variation.

The coefficient of variation is a kind of traffic light to signal how safe it is to use x to enforce comparisons with other groups or to make predictions about unobserved phenomena.

A green light gives you the right of way to use an intersection, but this right is no guarantee of safety. An errant driver traveling on a cross-street may run their red light and ruin your whole day. A coefficient of variation below 25% provides some assurance that we may use an average for comparisons or predictions. A low coefficient of variation layered on top of a robust sample size raises our level of comfort.

A coefficient of variation between 25% and 100% is analogous to a yellow light. You have the right to proceed for a limited

Table 5.2 Interpretation of the Coefficient of Variation (s/x)

Range of (s/x)	*Use of x for Comparisons or Predictions*
Up to 25%	Potentially safe
25% to 100%	Risky
Above 100%	Dangerous

time horizon; however, beware that drivers, bicyclists, and pedestrians are all waiting anxiously to jump into the intersection. Proceeding with a yellow light brings increased hazards.

A ratio of s/x above 100%, the equivalent of a red light, indicates that the standard deviation is larger than the sample mean. The high level of dispersion among data points shows that the mean is a dangerous measure of central tendency. Knowing x is better than not knowing x, but any decision to act solely on this figure is fraught with risk. You may face an emergency and have no choice but to run a red light; just recognize that this action may bring a wide range of outcomes.

Table 5.3 offers hypothetical data to give an example. Suppose the external relations office of a university surveyed graduates 10 years out to ask about annual pay. The analyst sorted responses by graduates of liberal arts and engineering courses of study to predict what current students could expect to earn in the future.

At first blush, it seems that engineers do better. A more numerate person would slow down and reflect on the relationships among n, x, and s in each group. We note that $n > 30$ for both samples, but we wonder how the data were collected. Our first task would be to ask questions to understand any steps taken to mitigate the collection of biased samples. For example, we may suffer from response bias, where only the most successful alumni chose to respond to the survey.

Next, we note that the value of s/x is much higher for liberal arts students than for engineers. A value of 93% for the former

Table 5.3 Annual Compensation at 10th Reunion

	Arts and Sciences	*Engineering*
n	220	145
x	$72,300	$91,005
s	$67,525	$33,000
s/x	93%	36%

group makes it borderline reckless to use x to make predictions or comparisons.

What's probably happening is that, by 10 years after graduation, engineering students have settled into a family of jobs with predictable compensation while liberal arts students have pursued more varied career paths. Jobs in this second group likely range from volunteer and nonprofit roles to positions in highly paid professions such as law, business administration, financial services, and medicine.

Given this data, I'd say to a current student who asked for advice,

Data collected from people who graduated 10 years ago suggest that engineering students consistently earn a good living. Graduates of the college of arts and sciences have significantly more varied compensation trajectories. Some make modest livings while others earn a lot of money. If this sample is representative of labor market trends, studying an engineering curriculum seems to be a less risky path if future pay is important to you.

A first cousin of coefficient of variation is ***volatility***, which measures variance over time. Interest rates, commodity prices, and foreign exchange values fluctuate daily. Unexpected changes in things that bounce around may cause problems. Organizations use risk management tools to measure and then cope with fluctuations of critical inputs beyond direct control. As with coefficient of variation, volatility is expressed as a number or percentage so that comparisons may be made across different samples.

Annualized volatility is an established way to measure how much something bounces around over time. Suppose a portfolio manager seeks to invest in the soft drink industry and is curious about recent stock price movement in Coca-Cola and Pepsi. A bouncy stock price history is an indicator of investor skittishness, a factor the portfolio manager deems to be of interest.

Table 5.4 Annualized Volatility of Peer Company Stocks

	Coca-Cola (KO)		*Pepsi (PEP)*	
Trade Date	*Closing Price*	*Daily Change*	*Closing Price*	*Daily Change*
11-Feb-19	$49.61	–	$112.97	–
12-Feb-19	49.66	0.1%	113.79	0.7%
13-Feb-19	49.79	0.3%	114.12	0.3%
14-Feb-19	45.59	–8.4%	112.59	–1.3%
15-Feb-19	45.24	–0.8%	115.91	2.9%
19-Feb-19	44.83	–0.9%	115.93	0.0%
20-Feb-19	45.10	0.6%	115.83	–0.1%
21-Feb-19	45.86	1.7%	116.10	0.2%
22-Feb-19	45.28	–1.3%	116.76	0.6%
25-Feb-19	44.94	–0.8%	116.06	–0.6%
26-Feb-19	44.69	–0.6%	115.97	–0.1%
27-Feb-19	44.94	0.6%	115.37	–0.5%
28-Feb-19	45.34	0.9%	115.64	0.2%
1-Mar-19	45.38	0.1%	116.18	0.5%
4-Mar-19	45.65	0.6%	116.17	0.0%
5-Mar-19	45.60	–0.1%	116.03	–0.1%
6-Mar-19	45.45	–0.3%	116.66	0.5%
7-Mar-19	45.28	–0.4%	116.10	–0.5%
8-Mar-19	44.84	–1.0%	115.23	–0.7%
	Standard deviation	2.1%		0.9%
	Annualization factor	15.9		15.9
	Annualized volatility	33.5%		14.0%

Table 5.4 shows one way of calculating volatility over 19 trading days in early 2019.

We begin by calculating daily changes in stock prices. The daily change is the current day's closing price divided by that of the preceding day and then subtracting 1 from the ratio (e.g., Pepsi's percentage price change on February 12 may be calculated as $113.79/$112.97 – 1.00 = +0.726%.

Some finance textbooks instruct readers to use the natural logarithm of the ratio of consecutive stock price observations (i.e., ln[$113.79/$112.97] = +0.723%). The two approaches yield almost identical results. My experience is that it is better to stay away from logarithms if you need to explain what you've done to a boss, board, client, attorney, or journalist.

Once we have daily percentage stock price changes, we calculate the standard deviation of these values in the sample. As noted earlier, the =STDEV.S function will do this in Excel. Notice how, for this sample of daily stock prices, Coke's standard deviation (2.1%) is twice that for Pepsi (0.9%). On a percentage basis, Coke's daily stock price strayed further from its mean during the sample period, suggesting a higher level of investor concern.

In the short run, Coke's higher volatility – at least over this thin sample – suggests that one should apply greater caution when using the average stock price to make a prediction. Noting that stock prices may not go below zero, some investors view elevated volatility not as a defect but as a feature because bouncier trajectories may propel stock prices to greater heights.

A convention for comparing volatility measures is to express them in annual terms. Some analysts use this number as the input for stock price volatility when using the Black-Scholes option pricing model.

The convention in industry is to annualize daily standard deviation through multiplication by the square root of 252 (the approximate number of trading days in a calendar year). If we had used, say, quarterly price changes, we'd multiply standard deviations of quarter price changes by the square root of 4 to convert them to annualized measures.

Our calculations show annualized volatility of 33.5% for Coke and 14.0% for Pepsi. Investors in Coke seem to have been much more nervous than investors in Pepsi over the sample period. I emphasize that 19 trading days represents a thin sample from which to make generalizations and that market conditions in early 2019 may not be representative of preceding and successive investing horizons.

I close this chapter by imploring you to never send or receive ratio data without ensuring that they are accompanied by measures for *n*, *x*, and *s*. Sometimes, though, this information is not available. Then we do the best we can using our wits and knowledge of sampling, average, and variance.

My vote for the greatest adjustment ever for a lack of information on *n*, *x*, and *s* goes to Stephen J. Gould, an anthropology professor on my college campus who railed against bad scientific thinking. After I had graduated, he was diagnosed with abdominal mesothelioma, a nasty form of cancer. His physician dodged questions about Gould's prognosis.

Gould headed straight to the medical school library to read the technical literature (2013). The brutal finding was that the disease was incurable and had a median mortality of eight months from the time of discovery. After 15 minutes of stunned silence, Gould smiled. Numeracy literally saved the day. There was hope.

First, Gould reasoned that median is a flawed measure of average. This tool parses data into halves surrounding the middle observation. The unlucky half of patients diagnosed with abdominal mesothelioma were expected to die within a few months; however, the rest were expected to live beyond the eight-month horizon.

Second, this statistic made no reference to the variance of the data points. Missing was any discussion of how far extreme values fell from the measure of central tendency. Further, the literature offered little information about the patients, the heterogeneity of people who comprised *n*.

Gould reasoned that the odds were stacked in his favor. He was young, benefited from an early diagnosis, had access to the best medical care in the world, and, most importantly, possessed an optimistic demeanor. He was confident that this collection of circumstances distinguished him from other patients in the sample and would propel him to the far-right side of the distribution. He concluded that he would be an outlier.

Using numeracy permitted Gould to make an astonishingly accurate prediction. He continued to enjoy a fulfilling life for the next *20 years*, publishing dozens of articles and books and enriching the lives of countless students. The eight-month median cited in the literature was, for Gould, a dysfunctional measure of central tendency.

Recap of Chapter 5 (Sample Standard Deviation [*s*])

- Variance, dispersion across data points, degrades the usefulness of averages.
- Standard deviation (s) measures the average distance of data points from their mean (x).
- Coefficient of variation (s/x) permits comparisons of variance across data sets.
- Numeracy rests on considering relationships among n, x, and s.

6

NORMAL DISTRIBUTION [$N(x,s)$]

This chapter combines x and s to introduce the normal distribution, a tool used to make predictions from samples. The notation $N(x,s)$ defines a normal distribution by pairing a measure for arithmetic mean with another for standard deviation. Thus, when I say $N(-3,4)$, I define a normal distribution with a mean of –3 units and a standard deviation of 4 units. Remember that x and s share the same unit of measure, but s is always positive.

Imagine you are one of 45 students filing in to my classroom. You take your seat and notice a penny sitting on your desk. I ask you to flip the coin 10 times, record the number of heads, and then repeat this process. We quickly obtain 90 observations.

I then go through the room, asking each student to call out the results of their two trials as I compile a histogram on the whiteboard. The x-axis shows the number of heads per trial, and the y-axis tallies the counts for the number of trials captured by each bin. Figure 6.1 shows the results of one such exercise. None of the 90 trials resulted in zero heads, a single trial had one head, five had two heads, and so on.

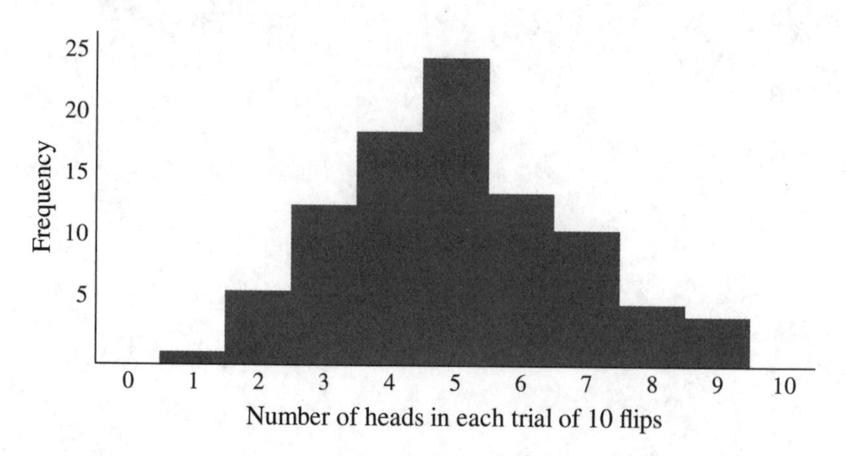

Figure 6.1 Distribution of Number of Heads per 10 Coin Flips (n = 90 trials)

Notice the distribution's shape. The most common outcome was five heads, the middle bin of this histogram, so we have a hump in the center. The number of heads per trial falls off as we move away from the middle toward each tail of the distribution.

Should we repeat the coin flipping process a really large number of times, the range of values on the *x*-axis would grow to include a few zeros and tens while the relative counts of bins across the *y*-axis would smooth out to approximate the outline of a symmetrical, bell-shaped curve.

The so-called ***normal distribution*** is the shape of a histogram that results from data pulled from many random, independent trials. For this book's purposes, the terms *bell curve* and *normal distribution* are used interchangeably. I emphasize that the bell curve is yet another idea, as opposed to a tangible thing in the real world.

While no single person may claim to be the inventor of this concept, an historian of statistics argues that German mathematician Johann Carl Friedrich Gauss (1777–1855) should get credit for putting forth a quantitative definition of the bell curve in the early 1800s (Stigler, 1986, 1999). This contribution was deemed to be so significant that the German government put Gauss's portrait plus an illustration of the normal distribution on the 10 Deutsche Mark currency note.

As an aside, Gauss had demonstrated exceptional math skills from a tender age. A story goes that one of his grade school teachers sought to keep him occupied by asking him to count the integers from 1 to 100. He instantly responded with the correct answer of 5,050. Item number 7 in Appendix B shows how the boy genius pulled this off.

The bell curve is useful in describing and predicting frequencies of phenomena involving many independent, random trials. I emphasize that there is probably no data set in the world that perfectly follows a normal distribution.

We define a normal distribution with an arithmetic mean and a standard deviation around that mean. Figure 6.2 shows a normal distribution for a data set for $x = 0$ and $s = 1$ or, more simply, $N(0,1)$. This particular curve is a special case because

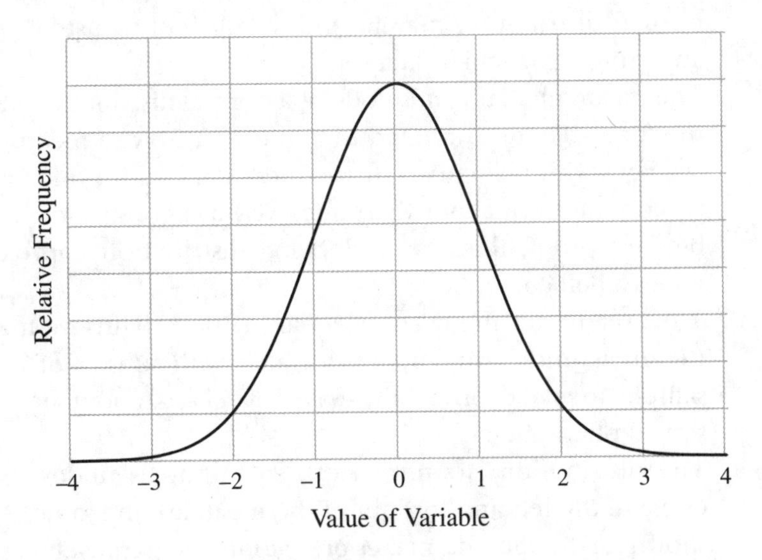

Figure 6.2 The Bell Curve of a Normal Distribution with $x = 0$ and $s = 1$

the hash marks on the x-axis measure both the number of units and the number of standard deviations from the mean. Notice how the downward-sloping curve looks like a bell. As we move one standard deviation from the mean, the curve hits inflection points and transitions from being concave down to concave up.

An interpretation of Figure 6.2 is that we have a histogram showing the distribution of counts of something we seek to measure. The most common occurrence is when the variable has a value of zero. We note the presence of fewer observations at –1 and +1 (a full unit below and above 0) and even fewer that occur at –2 and +2. Almost no observations appear at –4 or +4.

All bell curves share these six useful characteristics:

1. They are defined by two sample statistics, the mean (x) and standard deviation (s) of the underlying data. Thus, we only need these two statistics from a sample to create a normal distribution. The big warning, though, is that the normal distribution may be the wrong tool to use when analyzing a particular data set.
2. The mean, median, and mode of a normal distribution are the same. There's no need to argue over how to measure averages with bell curves. If we find a data set in which all three values are about the same, we have good evidence, but not proof, that the underlying distribution approximates a bell curve.
3. The area under the curve is equal to 100%. Thus we may use areas under the curve associated with subsets of the sample to make probability-based forecasts about unobserved phenomena.
4. The distribution is symmetrical, so the areas under the curve to the left and to right of the mean are the same. As shown further on, this makes prediction comparatively easy to do. We don't need to worry about the presence of especially large or small observations skewing our predictions.

5. The distribution has predictable tails. An exponential relationship between variance (distance from the mean) and frequency causes normal curves to drop quickly when observations stray from the middle. Notice how the curve gets insanely close – but never touches – the x-axis when the number of standard deviations from the mean falls below three or rises above three.
6. The distribution follows the ***empirical rule*** of having about 68%, 95%, and 99.7% of all observations contained, respectively, by subsets of the sample within plus-or-minus one, two, and three standard deviations in each direction from the mean. In this case, the standard deviation equals one unit, so the area under the curve from –2 to +2 envelops about 95% of all observations. Further, no more than about 0.3% (100% – 99.7%) of all observations in a normal distribution are larger than three standard deviations or smaller than three standard deviations from the mean.

Before showing how we may apply these rules, I emphasize that the normal distribution is not a single curve. Instead, there are an infinite number of curves that may be defined by a mean and standard deviation. Figure 6.3 compares three.

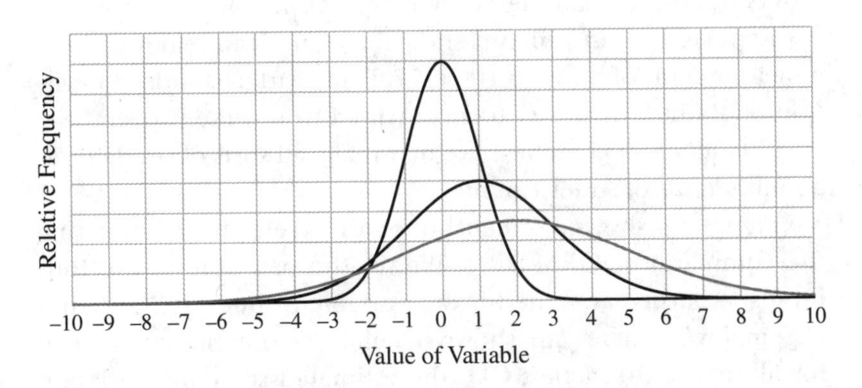

Figure 6.3 A Family of Normal Distributions

Curve A, the one with the highest peak, is the same distribution shown in Figure 6.2 with a mean of 0 and standard deviation of 1. Curve B ($x = 1$, $s = 2$) has the peak in the middle while Curve C ($x = 2$, $s = 3$) has the lightest shade.

Notice how varying x and s influence a bell curve's shape. A higher value of x pushes the curve to the right while a higher value of s flattens the peak. Curve A, with the lowest values of x and s, peaks on the left with more observations squished near the center. The lower variance causes the tails for Curve A to fall toward the x-axis the fastest, suggesting that observations far removed from their mean are less likely than for the other two distributions.

Using the empirical rule, we may say that about 95% of observations should fall within two standard deviations above or below the mean. When applied to Curve A, 95% of the observations are crammed into the comparatively narrow interval of [–2, +2], calculated as $x \pm 2(s)$ or $0 \pm 2(1)$. Using the same approach, we calculate a 95% confidence interval for Curve B as [–3, 5] and for Curve C as [–4, 8]. Note how Curve C, despite having the largest x, is endowed with the highest s and is thus more likely to have an observation equal to –4.

Bell curves become useful when we seek to make predictions about things for which we have observations that we believe to be independent. An example is the height of people.

Repeated measurements show that American males grow to a mean height of about 5′10″ (177.8 cm) with a standard deviation of about 3 inches (7.6 cm). Armed with these two statistics ($x = 70$ inches, $s = 3$ inches), we may make all sorts of predictions about heights of American men.

Figure 6.4 shows a screenshot of an Excel spreadsheet using the command =NORM.DIST. Within the parentheses are four inputs: a particular value for our variable of interest (say, adult U.S. males 70 inches tall, shown in cell F13), the estimated mean for all males (70 inches, C1), the estimated standard deviation for all measurements (3, C2), and TRUE, the instruction for the

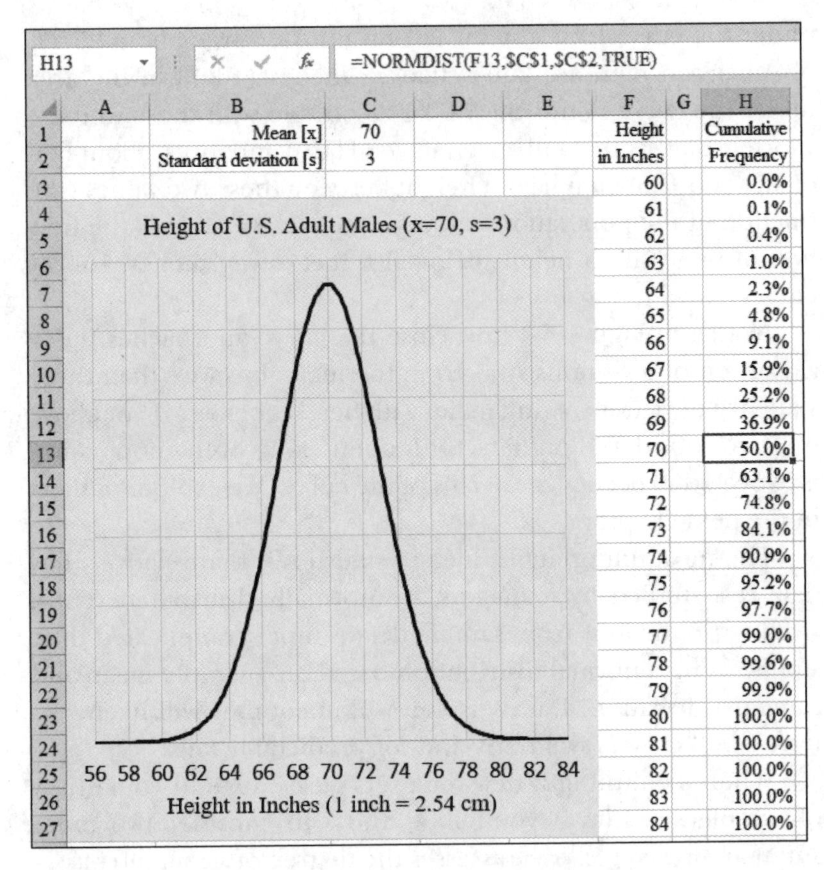

Height in Inches	Cumulative Frequency
60	0.0%
61	0.1%
62	0.4%
63	1.0%
64	2.3%
65	4.8%
66	9.1%
67	15.9%
68	25.2%
69	36.9%
70	50.0%
71	63.1%
72	74.8%
73	84.1%
74	90.9%
75	95.2%
76	97.7%
77	99.0%
78	99.6%
79	99.9%
80	100.0%
81	100.0%
82	100.0%
83	100.0%
84	100.0%

Figure 6.4 Using Normal Distributions to Predict Frequencies

software to display cumulative probabilities (the fraction of the area under the curve from the far left to a particular value for the variable of interest). In cell H13, we see that the area under the curve up to 70 inches equals 50%. Thus, we expect half of all adult males to be equal to or less than 70 inches tall.

The magic comes from our ability to use the percentages given in column H to predict how likely someone is to be below a given height, above a given height, or between two heights. For example, about 15.9% of adult males are shorter than 67 inches or 5′7″ (the value displayed in cell H10, which shows the area under the curve from the far left up until $x = 67$); about 25.2% of American men are taller than 72 inches (100% of the area under the curve minus the 74.8% associated with those who are shorter than six feet tall, given in cell H15); and, thus, about half of all adult U.S. men have a height between these two values (the fraction of the population below six feet, 74.8%, minus the fraction of men with a height below five feet, seven inches, 15.9%, equals 49.6%).

Notice in Figure 6.4 how close the curve approaches the *x*-axis when observations stray from the mean by more than three standard deviations. Adult males with heights below 5′1″ or above 6′7″ are considered outliers. Only about 0.3% of the population is expected to occupy these tails of the curve. We explore outliers in the next chapter.

The most undertaught idea in statistics is that relatively few data sets studied by managers are normally distributed. Thus, making predictions from sample data is more complicated than simply using a normal distribution based on a sample mean and standard deviation. Data sets you will encounter will likely not have a bell curve's symmetry and/or predictable tails.

When examining a new data set, please extend your study of sample statistics beyond n, x, and s to consider two more summary measures. ***Skew*** assesses the degree of asymmetry associated with a data set. In a normal distribution, high values offset low values so that the shape of the curve to the left of the mean mirrors that on the right. We are just as likely to be surprised by a big value as by a corresponding small one. Skewed data sets, by contrast, have a longer tail that points to one side or the other.

One may obtain a measure of skewness with the Excel command =SKEW. A value of 0 suggests a perfectly symmetrical distribution; a negative value suggests that outliers tend to be smaller values, relative to the mean; a positive value suggests that outliers tend to have larger values. Values below –1 or above +1 suggest worrisome skewness.

Suppose we manage a local film festival over a long weekend and sell tickets to specific movie showings. A colleague suggests we simply sell a more expensive movie pass and then allow patrons to watch as many movies as they wish. Figure 6.5 shows results of three imaginary scenarios for 200 customers. Each histogram shows the number of films viewed over the festival on the *x*-axis and the corresponding count of viewers who watched a given number of films on the *y*-axis.

Panel B of Figure 6.5 shows a perfectly symmetrical distribution centered around an average of five films. The measure of skew is 0 because the number of people who viewed six, seven, and eight films (or the median of 5 plus 1, 2, and 3) corresponds, respectively, with the number of viewers who watched four, three, and two films (or the median minus 1, 2, and 3).

In the absence of skew, the mean equals the median. The implication is that we're just as likely to have above- and below-average film watching frequency. We may reasonably assume that we have a normally distributed collection of patrons. Using our customary summary statistics ($n = 200$, $x = 5$, $s = 1$), we may infer that about 95% of viewers in our sample would watch between three and seven movies ($x \pm 2$ standard deviations). The likelihood of someone watching fewer than two or more than eight

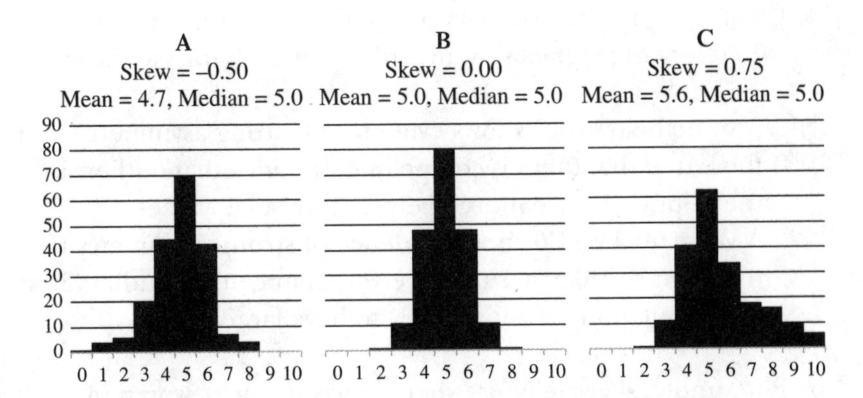

Figure 6.5 Skew Shows a Distribution's Asymmetry

movies ($x \pm 3s$) is remote (something like 100% – 99.7%). Relying on the simple math of a bell curve, I'd say, *Boss, I doubt that a patron would watch 10 films during the festival.*

Panel A shows a distribution with a skew of –0.50. The negative value means that the distribution is pulled farther to the left: a few people watched one film, a distance of four films below the middle value, while none watched nine films, a distance of four films above.

Evidence of negative skew is that the mean (4.7 movies per person) is smaller than the median of 5. We said in Chapter 4 that the arithmetic mean is more sensitive to the presence of outliers than the median is. Here, the presence of outliers on the left pulls down the mean more than the median. The implication of negative skew is that we're more likely to have outliers with a low value than an outlier with a large value.

Panel C of Figure 6.5 shows a distribution with a skew of +0.75. The positive value means that the distribution is pulled to the right: a number of people watched 10 films, a distance of five films above the middle value, while none watched zero films, a distance of five films below the middle value. Evidence of positive skew is that mean of 5.6 is greater than the median of five, showing how outliers have more influence on the mean than the median. The implication of positive skew is that we're more likely to be surprised by outliers with a high value than a low one.

When looking at data sets, my rules of thumb for skew are:

1. A value below –1.0 shows evidence of strong asymmetry to the left so that there is compelling evidence that outliers in the population are more likely to have small values.
2. A value above +1.0 shows evidence of strong asymmetry to the right so that there is a decent chance that outliers in the population are more likely to have large values.

Put simply, skew tells us whether we need to worry more about values for unobserved members of the population being especially small or large. Instead of giving quantitative probability distributions associated with a bell curve, we offer qualitative predictions. Thus, if shown data from Panel A, I'd say, *Boss, our sample shows a modest negative skew, which means that customers are a little more likely to watch a handful of films than to spend the weekend binge-watching movies. Offering a movie pass does not seem to make sense for our patrons.*

If shown the data from Panel C, I'd say, *Boss, our sample shows a meaningful positive skew, but the magnitude is not overwhelming. Customers are more likely to watch a bunch of films than just a couple. It makes sense to offer our patrons a movie pass option.*

When examining a new data set, after considering n, x, s, and skew, then I'd look at one more summary sample statistic. ***Kurtosis*** measures the thickness of tails, giving some idea of the likelihood of outliers. Whereas skew assesses whether outliers will be small or large, kurtosis assesses the chance that an extreme event could happen. As with skew, we use kurtosis to enforce comparisons with a normal distribution.

One may obtain a measure of kurtosis with the Excel command =KURT. A value near 0 suggests that the frequency of outliers falls at the rate that would be observed from a normal distribution (i.e., the probability of an observation straying from the mean by more than two standard deviations in either direction would be about 5% [100% – 95%] and by more than three would only be about 0.3% [100% – 99.7%]). A measure for kurtosis below –1 suggests that observations gravitate near the middle while a value above +1 suggests an elevated likelihood of extreme outcomes. Be careful when using other software packages, which may measure kurtosis using a different scale.

Figure 6.6 provides examples from additional imaginary scenarios for 200 moviegoers. Scenario B's distribution is mesokurtic, a fancy word that says that kurtosis is close to 0 and that the likelihood of extreme outliers is comparable to what we would find if we plugged the sample's values for x and s (in this instance, $x = 5$ and $s = 1$) into a normal distribution calculator. *Boss, our sample's kurtosis is 0, suggesting that maybe one or two patrons would watch 10 or more movies.*

Scenario A's distribution has negative kurtosis (–0.50), where the value below zero reflects the presence of thinner tails than would be displayed by a bell curve. This type of distribution is called platykurtic. I use the word *plateau* to remember this word because the top of the distribution looks relatively flat.

Evidence of thin tails is the absence of observations below 3 or above 7. The classic example of a platykurtic distribution is counting results from repeated rolls of a pair of dice: it is not possible to have outcomes below 2 or above 12. *Boss, our sample's kurtosis is –0.5, suggesting that it's unlikely that anyone will watch 10 or more movies. No one is going to be interested in a movie pass.*

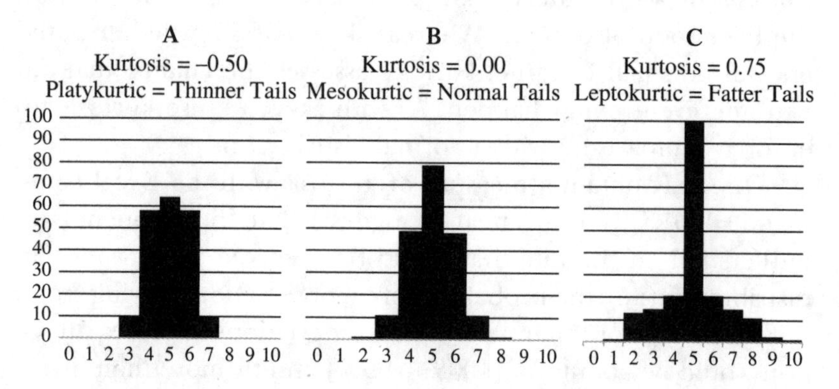

Figure 6.6 Kurtosis Shows Thickness of a Distribution's Tails

Scenario C's distribution has positive kurtosis, where elevated frequencies for values removed from zero reflect the presence of thicker tails than would be seen in a bell curve. This type of distribution is called leptokurtic, a word I remember through association with the word *leap*, which reflects the peaked density of observations near the middle.

The counterintuitive description of leptokurtic distribution may be understood by recognizing that the presence of thicker tails requires that there must be correspondingly fewer observations adjacent to the middle, making the central part look more peaked. If you were sculpting clay to make a volcano with an especially wide base, you would have to pull material away from areas near the middle. Put simply, positive kurtosis means fatter tails.

For scenario C, the substantial, positive value of +0.75 reflects the presence of observations as low as one and as high as nine. Outliers appear more frequently than what we would see if the distribution were normal. Stock market data sometimes display leptokurtic distributions because contagious investor greed or fear may bring sharp swings in security prices. *Boss, our sample's kurtosis is 0.75, suggesting that there will be some enthusiastic patrons who will watch 10 or more movies and that a few carpers will watch just a single film. Offering movie passes is a great idea.*

Measures of kurtosis below –1 suggest that extreme values should not be expected to occur, so one may take additional comfort in using measures of central tendency to make predictions or comparisons. Kurtosis values above +1 suggest that extreme values may eventually emerge, so one should probably shy from using measures of central tendency to make bold predictions or comparisons.

When using measures of skew and kurtosis, consider whether (1) *n* is fewer than 30 observations, (2) the sample may not be representative of the population, or (3) the sample is composed of heterogeneous observations. Of the three concerns, the thin-sample issue is perhaps the least worrisome. We discuss how to sort through this issue in Chapter 12.

There are times when we do not have access to the data associated with a sample. Instead, an analyst simply hands us summary statistics. If we make the assumption that the data are normally distributed, we may make probability-based forecasts armed with just x and s. As shown in Figure 6.4, we just plug these values into a normal distribution calculator such as the =NORMDIST command in Excel to create precise guesses for the likelihood that observations will fall above or below a value of interest or between two such numbers.

However, there may be times when we seek to make such predictions when we doubt that the data are normally distributed. If pressed to make a prediction under these circumstances, I fall back on a relationship between x and s identified by Russian mathematician Pafnuty Chebyshev (1821–1894).

He showed how a given share of observations must fall within a specified distance from the mean, regardless of how the data are distributed. His formula estimates the number of observations contained by k standard deviations from the mean, where k is greater than 1.

Chebyshev's theorem states that, for any distribution, the minimum fraction of observations that must fall within k standard deviations from the mean is equal to $(1 - 1/k^2)$. Commonly used distances are two and three standard deviations from the mean. Thus,

- For $k = 2$, at least 75% of observations fall within plus or minus two standard deviations from x.
- For $k = 3$, at least 89% of observations fall within plus or minus three standard deviations from x.

Notice how this theorem offers less-certain predictions. A confidence interval with a width of plus or minus two standard deviations will capture 95% of the observations for a bell curve but only 75% using Chebyshev.

Suppose we are handed the summary statistics presented in Figure 5.3 for annual compensation for the cohort of engineering students who graduated 10 years ago.

$$n = 145, \ x = \$91,005, \ s = \$33,000$$

Sam, an advisee, wants to know how likely it is that he could earn at least $200,000 per year a decade after graduation. We cannot evaluate skew or kurtosis of the individual observations, but we doubt that the data points follow a bell curve. We remember reading somewhere that most measures of human performance have a long tail.

Using Chebyshev's theorem, we may conclude that salary intervals for two and three standard deviations contain, respectively, at least 75% ($1-1/2^2$) and 89% ($1-1/3^2$) of the observations:

$$75\%:[\$25{,}005, \$157{,}005] = x \pm 2(s)$$
$$= \$91{,}005 \pm 2(\$33{,}000) \text{ for at least 108 students}$$

$$89\%:[\$0, \$190{,}005] = x \pm 3(s)$$
$$= \$91{,}005 \pm 3(\$33{,}000) \text{ for at least 129 students}$$

For the second confidence interval, I truncated the lower salary boundary at zero dollars, reasoning that one could not be paid a negative salary. Quick math suggests that at least 16 graduates (145 – 129) fall outside of the three-standard-deviation confidence interval and earn more than $190,000 per year. *Sam, my guess is that maybe 10% of engineering students who graduated a decade ago earn at least $190k. If you work hard, build professional relationships, and enjoy some luck, you have a shot at achieving your goal, but it's not a slam dunk.*

By inserting qualifying language and avoiding precise numbers, we demonstrate that we're numerate. We have extracted some information from a data set in a manner that allows us to be approximately right instead of precisely wrong.

Recap of Chapter 6 (Normal Distribution [$N(x, s)$])

- Normal distributions are defined by the mean [x] and standard deviations [s] of a data set.
- About 68%/95%/99.7% of data are contained within plus or minus one, two, or three standard deviations from x.
- Skew and kurtosis impair our ability to use the 68/95/99.7 rule.
- Chebyshev's theorem permits us to make less certain predictions from any type of distribution.

7

z-SCORE [z]

In the previous chapter, we introduced the ideas of skew (asymmetry of outliers) and kurtosis (likelihood of outliers). In this chapter, we reflect on the definition of outliers, their significance, and what we should do if we come across them.

We begin by presenting the symbol z, which signifies the ***z-score***, or distance a particular observation strays from a sample's mean, where the distance is measured in standard deviations. For the purposes of this book, we define z as:

$$z = \frac{(\text{Value of observation} - x)}{s}$$

To make this clearer, let's return to the compensation data for engineering alumni discussed in the previous chapter. The sample data are distributed such that $x = \$91{,}005$ and $s = \$33{,}000$. We look up the result for a particular alumnus and note that his reported annual pay was \$44,500. The *z*-score for this observation would be:

$$z = \frac{(\$44{,}500 - \$91{,}005)}{\$33{,}000} = -1.4$$

Remember that *s* shows how far, on average, a typical observation falls from the mean. In other words, standard deviation represents the typical distance an observation strays from $\bar{x}$. A value of \$44,500 has a *z*-score value of –1.4, indicating that this data point has a low value and is farther below the sample mean than that of the typical below-average observation.

Because a *z*-score's numerator and denominator share the same unit of measure, the resulting ratio offers a pure number that may be used to enforce comparisons across varied data sets. Once again, we have a measure that is an idea rather than a tangible thing. After a brief vignette, we'll discuss how *z*-scores may be used to identify outliers and then what to do should we come across them.

Undergraduate students learn to manage their time in order to accomplish academic, social, and extracurricular goals while in college. Through a trial-and-error process, students find ways to study effectively. My experience is that some students appear to be incredibly efficient, while their counterparts struggle to balance academics with social and extracurricular pursuits.

Figure 7.1 presents a graph for a collection of students in a particular class. The *x*-axis shows the number of hours each student had logged on to the course site within the university's learning management system (a *very* crude proxy for amount of time spent studying) while the *y*-axis shows the student's score on the mid-term examination (a proxy for the effectiveness of time spent studying). I caution that we have only one sample ($n = 1$), so it would be crazy to generalize from this stylized example.

Eyeballing the scatterplot shows that there appears to be a strong, positive, linear relationship between effort and

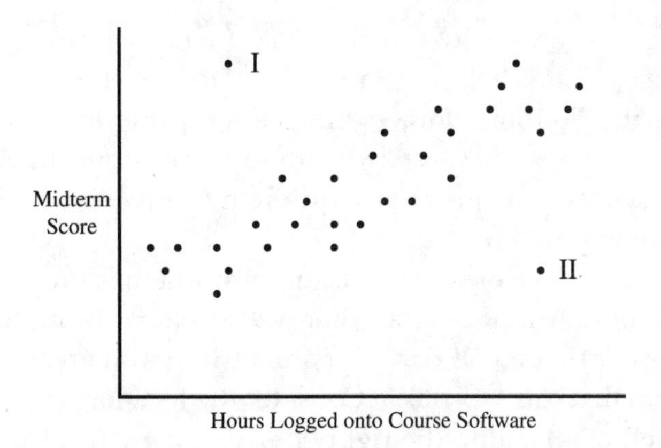

Figure 7.1 More Studying Generally Brings Better Grades

accomplishment. If we generalize from this sample, we could say to students that it pays to study. As a numerate person, you're likely to wonder how we could quantify relationships between studying and grades to make predictions for future classes. Discussions about correlation coefficient and slope, presented in future chapters, help us accomplish this goal.

Yet, we have two observations that stray from this relationship. Student I seems to have invested comparatively little time studying yet managed to crush the midterm. Student II put in the hours but failed to do well on the exam. As a teacher, I want to know more about these two students' experiences.

This small example highlights a disconnect between academics and practitioners (Andriani & McKelvey, 2007). Scholarly journal articles focus on central tendency (e.g., whether there is a relationship between studying and grades) while practitioner-focused management books concentrate on extreme performance (what we may learn from the experiences of Students I and II). Just reflect that business books displayed in airport bookstores describe spectacular successes and failures. Who wants to read a book about average companies?

My point is that we can't help but zero in on extreme values, which are the most interesting in a data set. When seeing lists showing, say, university rankings, CEO pay, or investment performance, I can't help but to first look to see who is at the top and on the bottom. Outliers are the most interesting observations in a data set because they are different from the majority of data points that cluster around the mean. We learn by studying extremes.

The use of *z*-scores helps distinguish whether the "tails" are unusual or downright extraordinary. Remember from the previous chapter that a bell curve is symmetrical, with an area under the curve that sums to 100%. Consider the locations of the curve to the left of –3 and to the right of +3 on Figure 6.2. Do you see how the curve seems to get insanely close – but never touches – the *x*-axis outside of these two reference points?

The empirical rule says that the area under the curve from below three to above three standard deviations captures 99.7% of all observations in this histogram. The areas below –3 or above +3 thus collectively represent just 0.3% (100% – 99.7%) of all observations. There is just a 1-in-333 chance of something appearing in one or the other of these two regions. Statisticians characterize observations with *z*-scores below –3 or above +3 as outliers. Public intellectual Malcolm Gladwell wrote a best selling book describing traits of people who are deemed to be outliers (2008).

Let's apply the *z*-score to three examples to help make the concept clearer. In Figure 6.4 we saw that adult U.S. males have an average height of about 5′10″ with a standard deviation of three inches. A *z*-score helps identify someone who would be considered remarkably tall. If we seek to solve for $z = +3$ when $x = 70$ and $s = 3$, then we arrive at a height of 6′7″ ($[79 - 70]/3 = +3$).

Someone who stands at 6′7″ comes across as being unusually, but not impossibly, tall. Assuming a normal distribution, only one in 666 adult males would be expected to be taller. We could come across someone who is seven feet tall ($z = +4.7$), but this would be an exceptionally rare occurrence outside of a basketball league.

In Figure 4.1, we showed results of a teacher's open-ended assignment asking students to read books over the summer break. Sample statistics for the class are $n = 26$, $x = 9$, $s = 5.5$. Lester appears to be the class oddball with 21 books completed last summer. His mother shares concerns with the teacher about whether he's an anomaly. The teacher calculates the *z*-score (([21 – 9]/5.5 = +2.2), notes that the value is comfortably within the boundaries of ±3, and shares, *Don't worry Mrs. Snow; Lester's reading level is admirable but not so unusual as to warrant special concern.*

My last example is from a thought experiment I offer in class. Suppose we will be asked to make separate trips to Phoenix, Arizona, and San Diego, California, at some point over the next year. I ask the students whether they would pack the same bag for each trip. Hot temperatures are little cause for concern, so we worry whether we should devote scarce packing space to bulky, cold-weather clothes.

Figure 7.2 presents average low temperatures for each city by month, obtained from Wikipedia articles for each city, and then a histogram of temperatures, assuming that they are normally distributed.

Notice how Phoenix has both a higher average (mean, median, and mode to the right of San Diego's) and a larger standard deviation (thicker tails).

One way to think about cold is the likelihood of snowfall. Snow starts to form when the temperature drops below 32° F (0° C). We could assess relative chances of snowfall by comparing the *z*-score for 32° within each distribution:

	Jan	*Feb*	*Mar*	*Apr*	*May*	*Jun*	*Jul*	*Aug*	*Sep*	*Oct*	*Nov*	*Dec*	*x*	*s*
Phoenix	45.6	48.7	53.5	60.2	69.4	77.7	83.5	82.7	76.9	64.8	52.7	44.8	**63.4**	**14.5**
San Diego	49.0	50.7	53.2	55.9	59.4	62.0	65.4	66.7	65.2	60.6	53.6	48.4	**57.5**	**6.6**

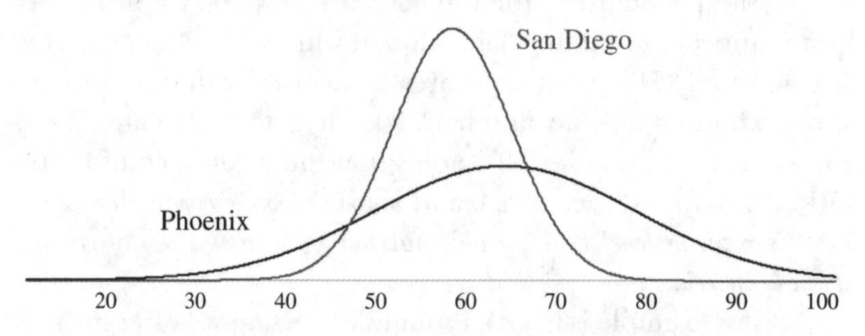

Figure 7.2 Average Low Temperature in Degrees Fahrenheit

Phoenix	*San Diego*
$\frac{(32° - 63.4°)}{14.5°} = -2.2$	$\frac{(32° - 57.5°)}{6.6°} = -3.9$

The probability of the temperature dropping to 32° in Phoenix is uncommon, as measured by a *z*-score of –2.2. Two standard deviations covers 95% of the observations for a bell curve, where half of those outlying observations (2.5%) would be expected to fall below –2. Phoenix's *z*-score of –2.2 makes this even less likely. An Internet search reveals that snow does fall in Phoenix from time to time, but it rarely sticks. The scope of Lester's summer book reading was about as infrequent – something of interest but not so freakish as to cause consternation.

As for snowing in San Diego, the *z*-score is –3.9, almost a –4. Relative frequency on a bell curve, as we move away from a mean, plummets toward the horizontal axis. Notice in Figure 7.2 that San Diego's gray curve drops insanely close to the *x*-axis once the temperature falls below the mid-30s. Having the *z*-score drop from –1 to –2 is a big deal, from –2 to –3 is a very big deal, and from –3 to –4 is an extremely big deal.

A *z*-score of –3.9 suggests that an event is incredibly unlikely, much like coming across a U.S. adult male with a height of about 58.3 inches (approximately 4′10″ or 1.48 m). A separate Internet search shows that snow in San Diego is incredibly rare. At the time of this writing, the last time snow flurries were seen was in 2008 and the last time snow touched the ground was in 1967, before most of my students were born.

Our study of *z*-scores suggests that we should consider the possibility of really cold weather as we prepare for a trip to Phoenix, but we need not bother with this contingency should we have to pack for a trip to San Diego.

The so-what is that any review of a new, unfamiliar data set should include taking time to calculate the *z*-score for each observation to look for the presence or absence of possible outliers, the most interesting of all observations. Admittedly, this is a fishing expedition because we don't know what, if anything, we'll find.

If we have data that appear to be normally distributed (e.g., produced from large number of independent trials), then observations with *z*-scores below –3 or +3 represent extreme values that should occur once every 333 or so observations.

Should we find observations with *z*-scores above +3 or below –3 on a more frequent basis, then we must use our wits to figure out how to interpret them. No software informs us of what outliers mean. I argue that there are three possibilities. The presence of more frequent than expected outliers indicates:

1. Something is going on;
2. We have problems with how we're measuring our variable of interest; or
3. There is evidence of interdependence among observations.

In Chapter 1, I introduced Claude Shannon, who said that information is a surprise. A message is informative if the chance of its occurrence is small. If a message is very predictable, then it has a small amount of information; no one is surprised to receive it. Seeing more frequent than expected outliers provides a clue that something surprising is going on in the world around us.

Suppose we have a factory that has hummed along for years without incident. The quality control team notices that the factory has recently produced units with a weight that is more than three standard deviations from its historical mean and that the incidence of overweight output is higher than the 1-in-333 rate implied by the empirical rule. These observations offer evidence that something is afoot.

Perhaps some unusual shock to our system like cold weather, worker illness, or a supply chain incident interfered with quality control measures at our factory. The presence of outliers alerts us that something has changed. This information is useful as we seek to bring our manufacturing process back under control.

A second implication of unexpectedly more frequent outliers is the emergence of a measurement problem, which is also information. Perhaps our quality control group changed people or processes so that we are no longer measuring tolerances as we had in the past. If this is the case, we need to revisit how we define and measure variables that interest us.

The third implication is that we are not dealing with independent trials that give rise to bell curves. Instead, there could be interaction among the things that we are measuring. If someone yells "Fire" at a movie theater, we don't expect the placement of people within a particular auditorium to be distributed normally across the room: harried customers will follow those who found the closest exit. Some type of feedback loop may have emerged in our factory so that the value of one observation may have somehow influenced the value of another observation.

Interdependence may give rise to a type of distribution identified by Italian economist Vilfredo Pareto (1848–1923). Like Gauss, Pareto tried to answer the question of how data are distributed in the world around us. Pareto is given credit for the so-called 80/20 rule where, say, 80% of land is owned by 20% of families, 80% of profits come from 20% of the products sold, 80% of points come from 20% of a team's players, and so on.

Pareto's ***power law distribution*** is often described by the formula $Y = X^{-a}$ where a is a constant that serves as an exponent with a negative value. Once again, we're working with an idea, not a tangible thing. I doubt that any phenomenon in the world perfectly follows a power law distribution. Both the bell curve and the power law distribution are histograms that display how frequently we expect to see different values of X, our variable of interest. With Pareto's distribution, Y gets smaller as X increases.

Figure 7.3 contrasts Pareto's distribution with Gauss's bell curve. The dark line shows a downward-sloping Pareto distribution that reflects ever-lower frequencies on the y-axis as the variable of interest increases on the x-axis. Small values of X are quite common while large values of X appear infrequently.

The important point about Pareto distributions is that extreme outcomes are uncommon but not impossible. As we move to the right on the x-axis, the tail of the dark curve is noticeably thicker than that for the gray, bell curve. Extreme events appear more frequently than would be seen under Gaussian distributions.

Normal distributions are associated with independent trials such as coin flips, dice rolls, or roulette wheel spins – one observation is not expected to influence the next. Power law distributions are associated with interdependent trials, where one

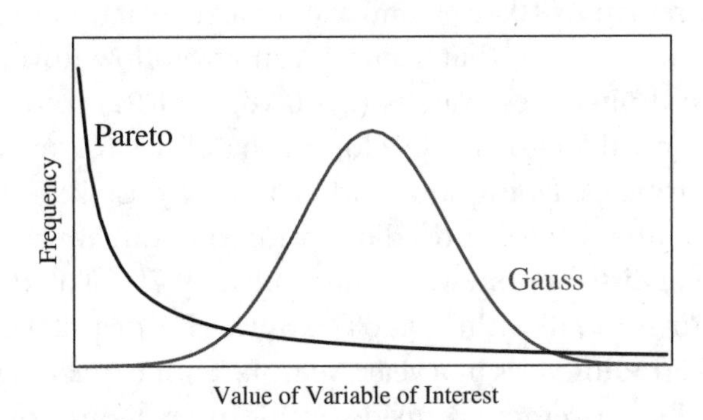

Figure 7.3 Pareto Distributions Have Fatter Tails

observation is expected to influence the next. Social interaction is interdependent.

Human behavior across most domains – for example, the number of times that a politician gets reelected, a professor gets published, an actress is nominated for an Emmy award, or an NBA basketball player scores points – displays frequency distributions closer to a power law than a bell curve (O'Boyle & Aguinis, 2012).

Put simply, when it comes to, say, playing basketball, the vast majority of us suck (as shown by a higher frequency of low point counts on the left) while a select few like LeBron James are able to consistently light up the scoreboard (as shown by a lower frequency of players posting high point counts on the right). We should be entertained but not shocked when outliers like LeBron join the ranks of professional basketball players.

Security returns in financial markets are interdependent. Investors watch each other and make buy or sell decisions based on how colleagues act. When there is a whiff of bad news, many investors sell first and ask questions later, much like moviegoers running for the exits if someone simply yells "Fire."

Consider the behavior of the U.S. stock market in 1987. I downloaded daily closing values for the Dow Jones Industrial Average from the beginning of January through the end of September. My data showed results for 188 trading days with an average daily return of +0.16%, and a standard deviation of 1.02%.

We could assume that security returns follow normal distributions and plug these values ($x = 0.16$, $s = 1.02$) into a normal distribution calculator to develop probabilities for varied levels of future returns. Doing so would be a huge mistake.

Pairing the three-standard-deviation rule with an assumption of a normal distribution, we would expect 99.7% of all daily stock price returns to fall within 0.16% ± 3(1.02%) or [–2.9%, 3.2%]. Given 252 trading days in a year, we would not expect to see daily stock market movements outside of this interval once over calendar year 1987.

However, on October 19, 1987, the Dow Jones Industrial Average lost 22.6%, generating a *z*-score of –22 ([–22.6 – 0.16)/1.02]. We noted earlier that a four-standard-deviation event, in a world assumed to follow bell-shaped curves, was equivalent to it snowing in San Diego. A 22-standard-deviation event would be impossible. We could trade stocks every day since the Big Bang and not expect to see a daily drop of this magnitude in 14 billion years.

But this event did happen, and stock market watchers saw equivalent stock market movements in other exchanges. The lesson is that interdependence brings extreme outcomes much more frequently than would be seen with bell curves. Consider this sentence written by Citibank executive Robert Rubin in his January 9, 2009, letter of resignation to the CEO as the global financial crisis unfolded:

> My great regret is that I and so many of us who have been involved for so long in this industry did not recognize the serious possibility of the extreme circumstances that the financial system faces today. (Rubin, 2009)

My interpretation is that Rubin and his colleagues made predictions of security prices based on models that used normal distributions. Doing so grossly underestimates the likelihood of market crashes. In *The End of Theory*, Richard Bookstaber (2017) shows how interdependence among people makes aggregate economic behavior unpredictable.

My favorite example of unpredictability following interdependence is when a 13-year-old Bobby Fischer, playing black in a chess match against master Donald Byrne, sacrificed his queen to pull off an inconceivable win.[1] Chess games are the ultimate example of interdependence, where there are countless

---

[1]https://en.wikipedia.org/wiki/The_Game_of_the_Century_(chess), retrieved 9 April 2021.

feedback loops as player 1 and player 2 try to anticipate each other's reactions to unfolding moves.

So, to summarize, when we see observations with *z*-scores below –3 or above +3, then we must work to rule out three hypotheses: something strange is happening, we have a measurement problem, or we're coping with interdependent data that may give rise to extreme values with a frequency that is somewhere below likely but above impossible.

Let me close this chapter with an anecdote to reinforce the idea that many things are not normally distributed. A senior member of our academic community joked at a faculty assembly that "75 percent of all MBA students think they're above average, but we all know that this is not possible." From now on, I hope the hair on the back of your neck stands up when you hear statements about the impossibility of extreme outcomes.

My highly educated colleague made the common mistake of visualizing distributions as symmetrical, with half of observations falling on either side of measures of central tendency.

Consider a scenario shown in Figure 7.4. An MBA course has 20 students. On a nice spring day, five of them skip class to go to a baseball game. Unfortunately, the instructor decides to give an unexpected quiz.

The 15 students who came to class scored an 80%, while the balance earned zeroes. The mean score was 60%, so 75% of the students earned above-average grades. In this case, the mean is well below the median, as evidenced by skew of –1.3, so a vast majority of students may be classified as above average.

1	*2*	*3*	*4*	*5*	*6*	*7*	*8*	*9*	*10*	*11*	*12*	*13*	*14*	*15*	*16*	*17*	*18*	*19*	*20*
0	0	0	0	0	80	80	80	80	80	80	80	80	80	80	80	80	80	80	80

Figure 7.4 Quiz Scores for an MBA Course (n = 20, x = 60, Median = 80, Skew = –1.3)

Recap of Chapter 7 (*z*-score [*z*])

- The *z*-score measures an observation's distance from the mean in standard deviations.
- Observations with *z*-scores below –3 or above +3 are often classified as outliers.
- Outliers indicate information, measurement error, or presence of interdependence.
- Distributions of human behavior often follow a Pareto distribution, not a bell curve.

8

CORRELATION COEFFICIENT [r]

In this chapter, I introduce r, a measure of ***correlation***, which assesses the degree to which movement in one variable is associated with movement in another. A more formal definition is that r assesses the strength and direction of a linear relationship between two variables. In Chapter 5, we used the tools of n and x to calculate s for one variable. Correlation uses a similar approach for measuring r across a pair of variables.

If two variables move up and down together, they are positively correlated; if one tends to go down while the other goes up, the two variables show negative correlation. Finding such relationships is a basis for prediction, the most important concept covered in this book.

I caution that correlation is like a chainsaw. In the hands of a competent user this tool is incredibly powerful; in the hands of an idiot, it may bring untold damage. The purpose of this chapter is to explain correlation and then lay a foundation for thinking about ways to use this tool productively. Using r properly requires judgment.

Let's return to the apple counts we encountered in Figure 5.1. As you may remember, on six separate occasions we opened the refrigerator's fruit drawer to count the number of apples on hand. In our six observations, we recorded one, two, three, four, six, and eight apples. Now, let's suppose that we also counted the number of oranges in the drawer during our apple inventorying. Associated with the apple counts were, respectively, four, five, seven, four, six, and 10 oranges. Figure 8.1 presents a visual representation of these ordered pairs.

The line shows a so-called best-fit line associated with these observations. The relationship between the dots and line illustrates the two basic ideas of correlation. The tightness of fit of dots around the line shows the strength of a linear relationship while the direction of the line (upward-sloping or downward-sloping) shows the sign of the relationship.

In this case the dots cluster reasonably closely around an upward-sloping line, so we may say that there is a strong, positive correlation between apple and orange counts. If we have conviction that this relationship will hold outside of the sample, we may use apple counts to make a prediction about the number of oranges. If our spouse tells us that there are few apples in the fridge, we may use the ***correlation coefficient*** to

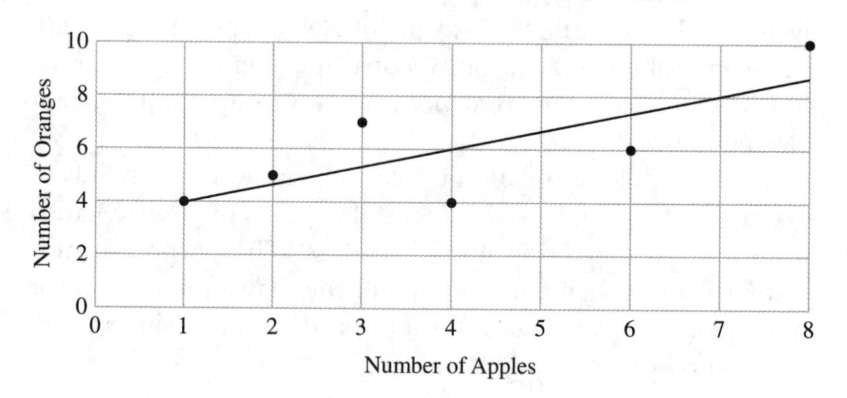

Figure 8.1 Apple and Orange Counts for Six Samples

support the prediction that we also likely have comparatively few oranges on hand.

The most commonly used measure of correlation is called Pearson's *r*, named in honor of English statistician Karl Pearson (1857–1936), but the person who actually developed the formula is a matter of debate (Stigler, 1986).

Pearson's *r* is used to measure correlation between two variables measured in ratio scales. Propeller heads have other tools to assess relationships among nominal, ordinal, or interval scales. In the pages that follow, we'll just stick with *r*.

The idea behind *r* is blending variance across two variables. In Figure 5.1, we calculated standard deviation by taking the average distance between a variable's observations and its mean. For correlation, we multiply these distances for two variables to assess the extent to which they move together:

- If there is consistently a large distance between observations and means for two variables, then we say that these two variables are positively correlated: especially large and small values for each variable tend to appear at the same time.
- If there are consistently large positive distances for one variable and large negative distances for the other, then we say that these two variables are negatively correlated. Large values for one variable tend to appear when small values are observed for the second variable.
- If there is an absence of a consistent relationship between distances in one variable with those of another, then we say that the two variables are not correlated. Large distances – positive or negative – in one variable are independent from large distances in the other variable.

Recall from Chapter 5 that we computed the standard deviation of the apple counts by measuring how far each observation strayed from the mean, squaring the differences to remove problems of aggregating positive and negative values, summing the squared differences, calculating the average distance (using $n - 1$ as the denominator to reflect uncertainty associated with using a sample), and taking the square root of the average distance to bring the units back from squared apples to just apples.

Table 8.1 shows how calculating r uses similar reasoning but multiplies differences across two variables. Please allow me a little space to unpack this table, which is less complicated than it first appears.

We start at the upper left. The lowercase i signifies each observation: $i = 1$ identifies the first observation, $i = 2$ the second, and so on all the way to the sixth and final observation. The labels X_i and Y_i signify, respectively, the number of apples and

Table 8.1 Calculating a Correlation Coefficient

	Apples	*Oranges*	*Mean values*		*(A)*	*(B)*	*(A × B)*
	X_i	Y_i	x_x	x_y	$(X_i - x_x)$	$(Y_i - x_y)$	$(X_i - x_x)(Y_i - x_y)$
$i = 1$	1	4	4	6	–3	–2	6
$i = 2$	2	5	4	6	–2	–1	2
$i = 3$	3	7	4	6	–1	1	–1
$i = 4$	4	4	4	6	0	–2	0
$i = 5$	6	6	4	6	2	0	0
$i = 6$	8	10	4	6	4	4	16
n	6	6				Sum of products	23
x	4.0	6.0					
s	2.6	2.3					

$$\text{Covariance}(X,Y) = \frac{\text{Sum of products}}{n-1} = \frac{23}{5} = 4.600$$

$$\text{Correlation}(X,Y) = \frac{\text{Covariance}(X,Y)}{(s_X)(s_Y)} = \frac{4.60}{5.95} = 0.774$$

oranges counted for a given observation. Thus, the third observation involved counting three apples and seven oranges.

The shaded area in the lower left presents the "big three" sample statistics for each variable: n (six observations for each variable), x (the arithmetic means of 4.0 apples and 6.0 oranges), and s (the sample standard deviations of 2.6 apples and 2.3 oranges).

The next two columns ("Mean values") simply place the means in rows to better show the calculation of differences in the following columns ("(A)" and "(B)"), which show the differences between observed apple and orange counts from their respective means. So, for the third observation ($i = 3$), the difference for apples is –1 (3 – 4), and the difference for oranges is +1 (7 – 6).

The final column ("(A × B)") multiplies the differences from the previous two columns. For the third observation, the product of the differences is –1. At the bottom of this column is the sum of products for all six multiplications, equal to 23.

If we have positive outliers occurring at the same time for each variable (i.e., the numbers of apples were large relative to the apple sample mean when the numbers of oranges were also large relative to the orange sample mean), then the sum of products becomes a large, positive value. If especially small apple counts occurred together with especially small orange counts, then the products of two sizable negative numbers would produce a large, positive sum of products because multiplying two negative numbers brings a positive value. Put simply, apple and orange counts moving up and down together create a sizable positive correlation.

If we have fruit counts moving in opposite ways (e.g., the numbers of apples were large relative to the apple sample mean at the same time that the numbers of oranges were small relative to the orange sample mean), then the sum of products would be a sizable negative number because multiplying a positive number by a negative number brings a negative value. This would be evidence of negative correlation.

A third possibility is that the apple and orange counts move independently, where there is no consistent relationship between especially large or small observations for each variable. In this case, positive and negative values in the final column cancel out so that the sum of products in the lower right part of the table would hover close to zero, which is evidence of an absence of correlation between the two variables.

There are two problems using the sum of products as a measure of correlation. First, it is a sum, so its value could be proportional to sample size: altering the number of observations changes the sum of products' magnitude, confounding our ability to use this number to enforce comparisons across other data sets.

A solution is to compute an average value. We do this in the row labeled "Covariance (X, Y)," where we divide the sum using the $n - 1$ convention ($6 - 1 = 5$) to reflect the uncertainty of working with a sample. The resulting ratio is called covariance. In this case, the value is 4.60. Covariance values may be compared across other data sets.

However, covariance is a completely impractical statistic. The unit of measure in this case is an "apple-orange" because we multiplied differences counted in apples against differences counted in oranges. My simple mind is incapable of visualizing an apple-orange. In a career using numbers, I never once worked with a colleague who used covariance to sort through a management problem.

The final step, to get rid of this bizarre unit of measure, is to divide the covariance by the standard deviation of each of the two variables. By dividing 23 apple-oranges by the standard deviations of 2.6 apples and 2.3 oranges, we arrive at a correlation coefficient of 0.774. Since the units of measure cancel out, the resulting statistic is a pure number. This step also brings the resulting ratio to between –1 and +1. The statistic r thus provides pure numbers with common scaling so that we may enforce comparisons over different pairs of variables.

There is no need to remember all this detail. Any statistical software package will calculate r for you. In Excel, one simply invokes the command =CORREL (data range 1, data range 2), so long as the two arrays have the same number of observations.

After you put away this book, just keep in mind that r has three useful properties. First, this ratio measures the strength and direction of a linear relationship between two variables. A positive value means that the two variables move up and down together while a negative value indicates movement in opposite directions. A larger distance from zero indicates a stronger relationship.

Second, the ratio is a unitless measure, so analysts may use this tool to make correlation comparisons across different pairs of variables. Finally, the measure always has a value between –1 and +1, so that the measure is easy to interpret. Figure 8.2 summarizes these points in tabular form.

Using this scale, we see that the 0.77 correlation coefficient indicates that, within our sample, there was a relatively strong positive linear relationship between number of apples and oranges in our refrigerator. Apple and orange counts tended to move up and down together in our sample.

If my classroom experience is any guide, it's at this point where people start to confuse correlation with the slope of the line connecting dots. Figure 8.3 disentangles the two ideas.

The five panels show plots of different samples of 21 observations so that the value of the X variable appears on the horizontal axis and the value of the Y variable appears on the vertical axis. The solid, best-fit lines represent attempts to connect each

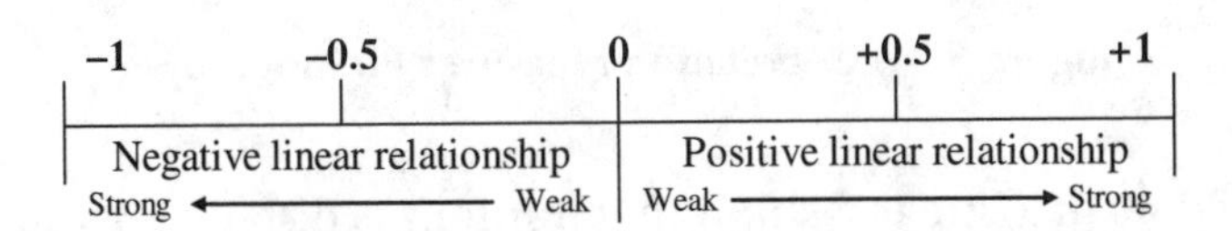

Figure 8.2 Correlation Measures the Direction and Strength of a Linear Relationship

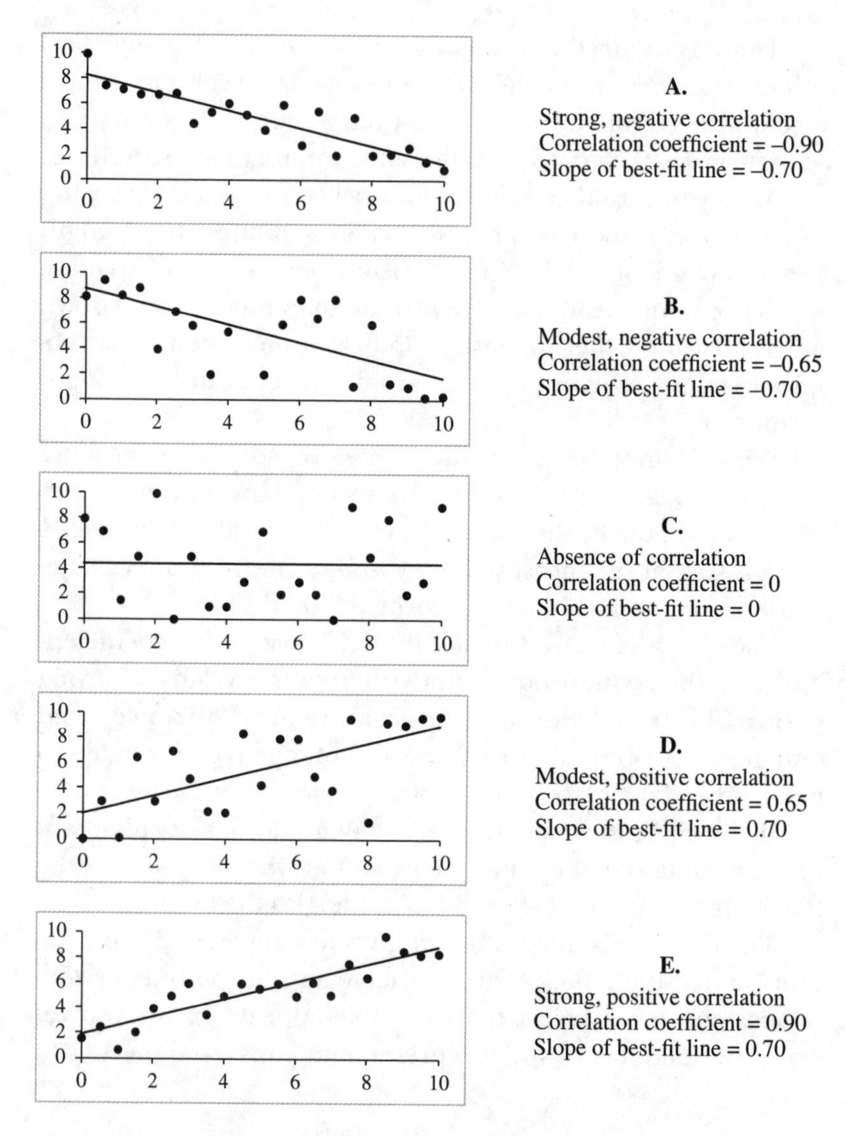

Figure 8.3 Correlation Measures Fit, not Slope

set of dots in a linear fashion that minimizes distances from individual data points.

Panel A presents a collection of observations with a strong, negative relationship. The correlation coefficient of –0.90 suggests that there is a tight inverse association between X and Y. When X goes up, Y almost always goes down. When X goes down, Y almost always goes up.

The slope of the best-fit line is –0.70, which says that, assuming there is a linear relationship between X and Y, an increase of one unit of X is associated with a 0.70-unit decrease in the value of Y.

The important point is to note that slope and correlation coefficient are different ideas. The slope of the best-fit line is how steeply it points down. The magnitude of the correlation coefficient is shown by the tightness of fit of dots around the best-fit line. The strong correlation is evident by the comparative lack of gaps between the dots and line.

Panel B presents a different collection of observations. The correlation coefficient of –0.65 indicates that there is a looser inverse relationship between X and Y. When X goes up, Y usually goes down. However, exceptions appear more often than in Panel A. The weaker correlation is evidenced by the presence of larger gaps between dots and the best-fit line.

Note that Panel B's best-fit line also has a –0.70 slope, suggesting an equivalent linear relationship between X and Y to the one shown in Panel A. This means that applying r to Panel B data would give the same prediction (an increase in one unit of X would give rise to a 0.7-unit decrease in Y), but we have comparatively less comfort than that the prediction will come true.

Panel C presents a special case: both the correlation coefficient and the slope are zero. Here there appears to be a complete absence of a linear relationship between values of X and Y. As X increases, Y may be just as likely to go up or go down. When r for a pair of variables is near 0, we have evidence to reject the idea that one variable may be used to make a prediction about the other.

Panel D shows a modest, positive correlation between X and Y. The moderate strength of the relationship ($r = +0.65$) is evidenced by the reasonably good clustering of dots around the upward-sloping best-fit line. The line's slope of +0.70 suggests that a one-unit increase in X is associated with a 0.70-unit increase in Y.

Panel E shows a strong, positive correlation between X and Y. The substantial strength of the relationship ($r = +0.90$) is shown by the tight fit of dots near the best-fit line. The slope of the line is also +0.70, which means that analyses of r for data sets in Panels D and E would give rise to the same prediction, but we would take greater comfort in our guess if it were based on the sample shown in Panel E.

Correlation matters because it is the foundation of prediction. Entrepreneur and polymath Jeff Hawkins argues that prediction is not just one of the things your brain does but the primary function of the neocortex and foundation of intelligence (Hawkins & Blakeslee, 2004). If you want to be a rock star at your organization, then develop the reputation of consistently making accurate predictions. Organizations that are endowed with the ability to make accurate predictions are much better positioned to experiment and innovate.

An example comes from my experience working for a car insurer that sold policies to price-sensitive motorists. Firms that charge too much fail to grow because new customers flock to competitors offering lower prices. Insurers who don't charge enough go broke because accident costs overwhelm inadequate premium collections. The trick is to predict the cost of insuring a given driver and then making sure that the quoted price is high enough to cover future accident costs and related expenses but low enough to attract shoppers.

One path to success is to build models that yield accurate predictions of future accidents – or at least do a better job of it than the competition. My employer invested heavily in systems and people who collected detailed data about loss costs and observable traits about the people who drive the vehicles. Using correlation analysis, we were able to find patterns that were helpful in setting insurance prices.

To take one example, the conventional wisdom in the industry was that size of the engine for a motorcycle was the best predictor of future accident costs. An enterprising product manager collected data on loss costs, engine size, and driver age. Using correlation analysis, he found that the age of the driver was a better predictor of accidents than the size of the motorcycle's engine.

The issue he faced was deciding whether the experience shown in historical data would recur in future samples. He concluded that the pattern found was a trend, not a blip, and then set prices based on distinctions of both youthful and mature drivers as well as small and large engines. The rates for mature drivers were set lower than those for the competition across both engine sizes. My firm was able to offer competitive prices for mature drivers and still make a profit.

A critical point to consider is that correlation analysis does not answer the question of why the pattern exists. Why are youthful drivers worse than mature drivers? I don't know, and I'm not going there. The last thing I want to do is to get up in front of a group of motorists, journalists, regulators, or agents and try having a calm conversation about the influence of age differences among people in society. Nothing good would come out of this: I would likely irritate audience members, attract unwanted publicity, and perhaps scare off potential customers.

What really matters is whether the relationship holds out of sample. A sensible course of action is to lower prices for mature motorists, raise them for youthful drivers, and then monitor results. If the prediction proves to be accurate, then the firm collects higher premiums for riskier young drivers (or motivates them to buy insurance from competitors who charge inadequate rates) and attracts more mature drivers at appropriate prices. Until competitors caught on, my employer experienced profitable growth while competitors charging age-neutral prices suffered underwriting losses resulting from failing to charge enough for youthful drivers.

Correlation helps reduce uncertainty without requiring the user to understand causal relationships among variables. The challenge is guessing whether the observed relationship will hold for unobserved data outside of the sample. Mastery of this basic idea helped make the CEO of our firm a billionaire.

Before closing out this chapter, let me offer three warnings about the use of r in making management decisions. Correlation is a helpful but not a foolproof means of putting forth predictions.

The first problem is that Pearson's r, the traditional measure of correlation, looks for the presence of a linear relationship between two variables. As one variable chugs along (i.e., grows arithmetically), another may grow at an accelerating rate (i.e., investments benefiting from compound interest over time) or decelerating rate (i.e., benefits from ever-increasing levels of employee training). Any use of r to evaluate nonlinear relationships may be misleading. We'll revisit this problem in the next chapter.

The second problem is ***spurious correlation***, where we find a strong linear relationship between two variables when the results are simply due to chance. An Internet search reveals many examples. One of my favorites is that over the period 1999 through 2009, the correlation coefficient between annual U.S. spending on science (as measured in billions of dollars) and number of suicides per year by hanging was an amazingly high 99.8%.[1] I do not believe that the implication of this statistic is to reduce societal investment in science to save lives. This outcome was likely a fluke that should be ignored.

The final problem is that correlation tests are designed for ratio data. Should your observations include nominal, ordinal, or interval scales, things could get tricky. Begin your analysis with Pearson's r to search for patterns but take the important step of consulting with a propeller head should you find anything of interest.

---

[1]Obtained from http://www.tylervigen.com/spurious-correlations on 24 April 2019.

Recap of Chapter 8 (Correlation Coefficient [*r*])

- Correlation is a basis for making predictions, not a causal explanation.
- Pearson's *r* assesses the strength and direction of a linear relationship.
- Correlation measures tightness of fit of dots around a best-fit line, not its slope.
- Nonlinear relationships and spurious relationships confound the ability to use *r*.
- Consult with an expert if your data have nominal, ordinal, or interval scales.

9

COEFFICIENT OF DETERMINATION [r^2]

Chapter 8 introduced r, which assesses the strength and direction of a linear relationship between two variables. A strong value of r may allow us to use observed movement in one variable to make predictions about the movement in another, unobserved variable. Intelligent use of this simple idea has allowed my old employer to flourish in a highly competitive industry.

This chapter introduces the idea of squaring r (that is, r^2) to assess how well movement in one variable fails to explain movement in another. Put another way, once given r, one may use some quick mental math to rule out a possible hypothesis for making predictions and explanations for the world around us.

Before going into detail, let's take a brief detour to introduce the idea of ***falsification***, which is a process of showing that something cannot be true. The modern proponent of this idea was Karl Popper (1902–1994), who argued that we cannot prove hypotheses but only reject them (1959).

Suppose we grew up in Europe and have seen only white swans. We could have a robust sample, say $n > 500$ observed swans, and note without fail that every swan we have seen is white. We could make an inference from this sample that the population of all swans on the planet are white.

Have we proved that all swans are white? Nope. There is always the chance that an unseen black swan lurks around the next corner. Expanding our sample size from 500 to 1,000 swans would offer additional assurance that our inference is reasonable. However, continuing to increase the sample size never proves that all swans are white. In the late 1690s, naturalists observed black swans in Australia, allowing European-based observers to falsify the idea that all swans are white.

Even though we may only prove things wrong, I support many initiatives that fall short of complete proof. Engineers have models that predict and explain how airfoils move through the atmosphere. None of these models has ever been proven to be always and everywhere true. I accept the vast, accumulated experience that goes into designing airliners and choose to entrust my life to commercial air travel. The use of pragmatism to reach a conclusion is discussed in Chapter 17.

Now, let's go back to the fruit drawer introduced earlier in the book. We found in Figure 8.2 that the correlation coefficient between apple and orange counts is 0.77. The statistic indicates that the number of oranges tends to rise and fall as does the number of apples. The purpose of this chapter is to provide a means of assessing how much comfort we may take in making predictions from this observed relationship.

Our path forward is to square the correlation coefficient ($r \times r = r^2$) to provide the ***coefficient of determination.*** This measure, with a value falling between 0 and 1, indicates the degree to which movement in one variable may be explained by a linear relationship with another.

A measure of 0 suggests that none of the movement in one variable may be explained by movement in another, while a value of 1 suggests that all the movement in the first variable may be explained by movement in the other. In real-world data sets, the value almost always falls between these two extremes.

In Table 8.1, we saw that the correlation coefficient for six observations in our refrigerator's fruit drawer was 0.774. Squaring the correlation coefficient brings a coefficient of determination of about 60% ($0.774^2 = 59.8\%$). This suggests that 60% in the variation in the counts of one fruit may be explained by variation in counts of the other. More importantly, about 40% of the movement in the number of oranges cannot be explained by movement in apple counts. Figure 9.1 presents, in three panels, why this is the case.

We start in the upper left with Panel A. The graph repurposes Figure 8.1, plotting the six observations of fruit counts but adds a formula for the best-fit line that connects the dots. Derived from a linear regression, a topic covered later in this book, the equation facilitates predicting orange counts given the

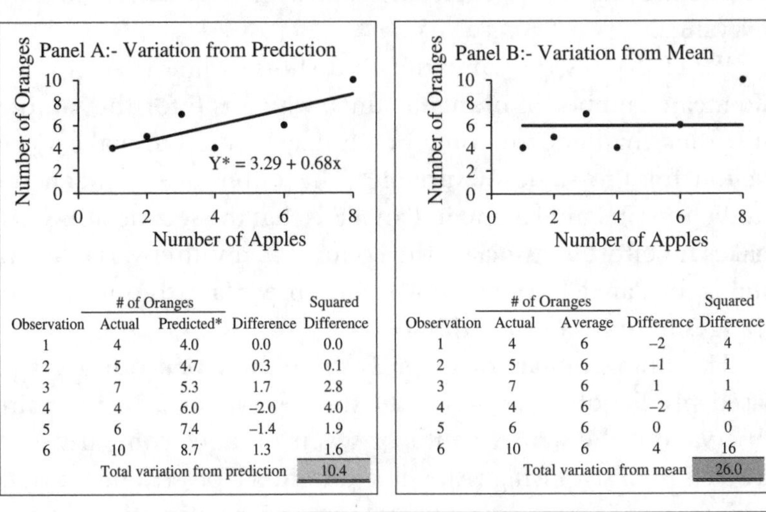

Observation	# of Oranges Actual	Predicted*	Difference	Squared Difference
1	4	4.0	0.0	0.0
2	5	4.7	0.3	0.1
3	7	5.3	1.7	2.8
4	4	6.0	−2.0	4.0
5	6	7.4	−1.4	1.9
6	10	8.7	1.3	1.6
	Total variation from prediction			10.4

Observation	# of Oranges Actual	Average	Difference	Squared Difference
1	4	6	−2	4
2	5	6	−1	1
3	7	6	1	1
4	4	6	−2	4
5	6	6	0	0
6	10	6	4	16
	Total variation from mean			26.0

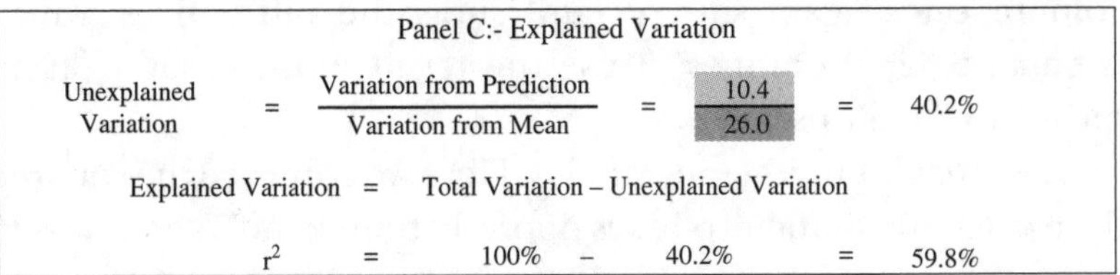

Panel C:- Explained Variation
Unexplained Variation = Variation from Prediction / Variation from Mean = 10.4 / 26.0 = 40.2%
Explained Variation = Total Variation – Unexplained Variation
r2 = 100% – 40.2% = 59.8%

Figure 9.1 r^2 Is a Measure of Variation Explained by a Linear Relationship

number of observed apples. A guess for the number of oranges may be calculated as 3.29 plus the number of apples multiplied by 0.68. Thus, the third observation with 3 observed apples has a predicted orange count of 5.3 = 3.29 + (0.68) (3).

The bottom chart of Panel A shows how well our model fares across all six observations. In the third sample, there were 3 apples, a prediction of 5.3 oranges, and an actual count of 7 oranges. The difference between the actual and predicted values was +1.7 oranges. The next column squares the differences to eliminate the problem of offsetting positive and negative values. At the bottom is the sum of squared differences, measured as 10.4 oranges^2. This panel offers one way of assessing the total variation of actual orange counts from predicted orange counts. In other words, this is an assessment of cumulative prediction error. I use the term "Total variation from prediction" to label this value.

Panel B plots the same dots but shows a line that captures the mean number of orange counts, which is 6 for the sample. In earlier chapters, we noted how x may be used to make a prediction for the value of variable. The difference between the predictions in Panel A and in Panel B is that the second approach makes no effort to associate the count with another variable. Put simply, in Panel B we are not trying to explain the number of oranges in terms of something else.

The bottom chart of Panel B shows how well our average-based prediction does across all six observations. In the third observation, the actual count of seven oranges compares to a prediction of six, giving rising to a difference of +1. The far-right column once again squares and sums the differences, which is equal to 26.0 oranges^2. I use the term "Total variation from mean" to label this value.

The magic emerges in Panel C. Here we compare the squared differences from the two bases of prediction. Panel A attempts to predict the number of oranges based on the number of observed apples while Panel B does the same with reference to only the mean of orange counts. If Panel A has smaller differences, then we have evidence that using the relationship between apples and oranges improves our ability to predict the number of oranges. If the differences are equivalent, then we may conclude that apple counts are not helpful in forecasting orange counts.

The sum of differences in the apple-based predictions is 10.4 oranges^2, considerably smaller than the differences in the sum of orange-based predictions of 26.0 oranges^2. Dividing the 10.4 by 26.0 gives a ratio of 40.2%, the share of differences in the oranges-based prediction that remains unexplained after invoking an apples-based prediction. Note that the 40.2% figure is a pure number because the units of measure in the numerator and denominator canceled out. I use the label "Unexplained variation" to describe this ratio.

The final step is to define the term "Explained variation" as the result of Total variation less Unexplained variation:

$$\text{Total variation} - \text{Unexplained variation} = \text{Explained variation}$$

$$100.0\% - 40.2\% = 59.8\%$$

Remember from Table 8.1 that the correlation coefficient for apples and oranges was 0.774. Squaring the correlation coefficient gives us the explained variation ($0.774^2 = 59.8\%$). Movement in apple counts explains about 60% of movement in orange counts.

By now, your head is probably spinning. Here is all you need to remember:

1. r assesses the strength and direction of a linear relationship between two variables, where the measure is a pure number that ranges from –1 to +1;
2. r^2 shows the degree to which movement in one variable may be used to explain movement in another, where the measure is a pure number that extends from 0 to 100%; and
3. $1 - r^2$ shows how much of movement in one number cannot be used to explain movement in another using a linear relationship.

As you were likely taught in previous statistics classes, correlation does not imply causation. There is no basis for saying that changes in Y are the consequence of changes in X just because X and Y have a meaningful correlation coefficient.

In the immediate case, no one is arguing that the number of apples is the cause for the number oranges. What we can say is that measurement of r reveals a value of 0.774, suggesting a strong positive linear relationship between these two variables. The resulting value of r^2 of 59.8% means that much of the movement in orange counts may be explained by movement in apple counts. A likely causal explanation is that changes in counts of both fruits reflect the recency of trips to the grocery store.

The more important measure, though, is $1 - r^2$, which shows that about 40% of the movement in orange counts *cannot* be explained by changes in apple counts. We have evidence that something other than consideration of the number of apples is needed to predict or explain the number of oranges. Perhaps what's missing is considering the number of people in the household or variation in their eating habits over different samples.

Here, then, is one of the most important ideas in this book: the term $1 - r^2$ lays the foundation for falsifying hypotheses developed from correlation analysis. We spend more time discussing falsification in Chapter 13.

Let's try another example to attempt to make this clear. Suppose we plan to move and wish to predict the selling price of our current home. Using an online tool, we collect data on six recent home price sales in our neighborhood. We've heard that realtors use house size, as measured in square feet, to value homes. Happily, our online tool also lists the square footage of homes sold.

Figure 9.2 presents data collected in an Excel worksheet. Column C shows house size in square feet, and column D shows sales price in dollars for the six houses in our sample. Cell D10 shows the calculation for r, D11, for r^2; and D12, for $1 - r^2$.

Using what we've learned, we may now make three conclusions from our thin sample. First, the measure of $r = +0.837$ suggests

D10 =CORREL(C3:C8,D3:D8)

	A	B	C	D	E
1			**House Size**	**Sales Price**	
2			Square feet	U.S. Dollars	
3			1,800	239,000	
4			3,800	364,000	
5			3,000	328,000	
6			2,750	315,000	
7			4,100	512,000	
8			3,450	500,000	
9					
10			Correlation coefficient [r]	0.837	
11			Coefficient of determination [r^2]	70.0%	
12			Unexplained variation [1– r^2]	30.0%	

Figure 9.2 Use of 1 – r^2 to Assess an Explanation
Source: Used with permission from Microsoft Corporation.

that there is a strong positive, linear relationship between square feet and sales price. We are comfortable saying larger homes command higher sales prices in our neighborhood. The positive correlation also means that these two variables move down together, so that a smaller house will garner a more modest sales price. Note that r does not indicate how much sales price changes for a given increase or decrease in square feet.

Second, we note that r^2 equals 70%. This relatively strong measure of coefficient of determination says that seven-tenths of movement in sales prices may be explained by movement in square feet. If this value were 0%, we would say that house size is not at all useful in predicting home prices, while a 100% value would mean that size could perfectly predict sales prices.

Third, we see that the value of 1 – r^2 is 30%, suggesting that about a third of movement in home sales prices cannot be explained by changes in home sizes. One or more other factors are at play when a buyer and seller agree on a price. Possibilities may include the age of the home, use of brick finishing on the facade, or quality of the landscaping. We would need to gather other data to test these hypotheses.

Thus the size of the house seems to be a pretty good basis for explaining past home prices. We do not yet have the tools needed to make predictions about the price that our own home will command, a topic we'll cover in Chapter 15.

To take a third example, suppose that an executive wonders about the relationship between employees' academic records and performance evaluations after the first year. Recruiting is expensive, and the firm wishes to get the most from its recruiting budget. The executive hypothesizes that a candidate's grade point average could be an indicator of subsequent employee performance and asks an analyst to test the hypothesis.

The analyst obtains high school GPA, as reported on the application form, and first-year performance evaluation score, as kept in the employee personnel files, for 275 people. Many files had missing GPA records and/or missing performance evaluations. The analyst converted reported GPAs to a uniform four-point scale and converted evaluations expressed as "does not meet expectations/meets expectations/exceeds expectations" to a three-point scale.

The analyst reports back to the boss with the following calculations: $r = -0.40$, $r^2 = 16\%$, and $1 - r^2 = 84\%$. Based on what we've covered in this book so far, all sorts of alarms should be going off in your head. A responsible analyst would say something like:

Boss, my quick analysis suggests that the hypothesis does not stand up to scrutiny. A review of 275 employee files revealed a weak, negative relationship between grades and performance evaluations; that is, the correlation coefficient is going the wrong way. Further, changes in GPA do not explain 84% of variance in performance evaluations, so I think grades are a lousy predictor of performance.

However, we have other problems. I was only able to use employees who had both a reported GPA on the application form and a one-year performance evaluation in their personnel file. This means we may have a biased sample, which risks bad inferences.

Finally, I had to convert ordinal measures of grades and performance evaluations into ratio scales to complete my calculations, so I'm worried that the transformed variables do not measure what we seek to assess.

Personally, I'd drop the study. If you want me to pursue it, then I ask to bring in someone from human resources who can get us additional data to expand the sample and a statistician to help us refine the methodology.

The point of this vignette is to show how a simple application of *r* may be used to rule out a potentially stupid idea. A modest investment in time together with a very basic understanding of statistical reasoning permitted us to quickly drop an idea that appears to have little merit. If we fail to persuade our boss, then at least we have a basis for having an intelligent initial conversation with subject matter experts brought into the project.

Before closing this chapter, let me offer one more cautionary warning about using r^2 to assess explanations or predictions. Consider Figure 9.3, which graphs a nonlinear relationship: here, the dependent variable grows with the square of the independent variable ($Y = X^2$) over the domain from 1 to 10.

The correlation coefficient of $r = +0.963$ suggests an extremely strong positive, linear relationship between the values of X and

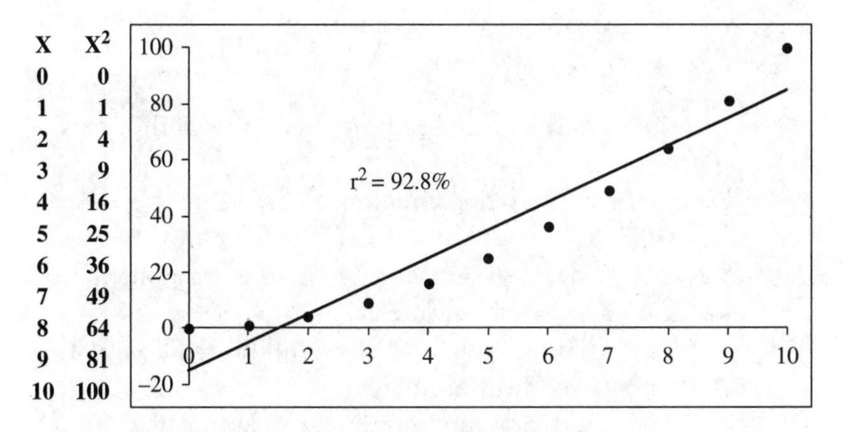

X	X^2
0	0
1	1
2	4
3	9
4	16
5	25
6	36
7	49
8	64
9	81
10	100

Figure 9.3 A Strong r^2 May Be Misleading

Y for the 11 dots. The strength of the correlation coefficient is revealed by how tightly the observations hug the best-fit line drawn among the dots. The impressive 92.8% value of r^2 shows that almost all the movement in the dependent variable could be explained by movement in the independent variable. The complement of this number, $1 - r^2 = 7.2\%$, suggests that only about 7% of the behavior of Y cannot be explained by a linear relationship with X.

Put simply, our analysis could give us confidence for making predictions for Y from yet-to-be observed values of X. Sadly, these predictions could be horribly wrong. Our best-fit line would be useless for making predictions if X were, say, –10 or +20. Predicting a Y from values of X that are well outside observed samples is known as ruthless extrapolation.

My experience with using r is that it is often possible to draw a straight line through observations that do not follow a simple, linear relationship and have statistical evidence of a decent "fit." Thus, any conclusions reached from statistical reasoning should always be subjected to subsequent testing using new data sets. Conclusions reached from statistical analysis are always provisional.

Recap of Chapter 9 (Coefficient of Determination [r^2])

- Hypotheses may be ruled out more easily than they may be proven true.
- The coefficient of determination, r^2, has values between 0 and 100%.
- r^2 assesses how well movement in one variable may explain movement in another.
- $1 - r^2$ assesses our ability to falsify hypotheses about correlation between two variables.
- Over a limited domain, nonlinear relationships may show strong values for r and r^2.

10

POPULATION MEAN [μ]

We now introduce μ, the Greek letter mu, which signifies the arithmetic mean for all data points in a population. To use a fancy word, μ subsumes $\bar{x}$. The value of $\bar{x}$ depends on the sample taken from a population, while there is generally only one μ for a population at a given point in time. In many domains, measuring $\bar{x}$ with certainty is relatively easy but doing so for μ is impossible.

Our discussion so far has shown how n, $\bar{x}$, and s are building blocks used to measure z, r, and r^2. These six measures form the basis of ***descriptive statistics***, a set of tools that help us find quantitative ways to summarize historical data sets.

The balance of this book extends our work to cover ***inferential statistics***, which means drawing provisional conclusions about unobserved phenomena based on relationships identified from observed data. Doing so helps us work toward Nobel laureate Milton Friedman's ultimate goal for science of developing meaningful predictions about things that have not yet been observed (1953). Inference is the act of working through the bottom arrow in Figure 3.1.

We are rarely able to measure populations. Consider the problem faced by a marketer seeking to understand the buying behavior of people living in Asia. The population is enormous, and it would be incredibly expensive to survey all people within the population. As time progresses through a comprehensive survey process, the population's composition and buying behaviors change.

The best we can do is to measure a variable of interest in a sample and then draw an uncertain inference about that variable within a broader, unobservable population. The good news is that statistical inference has repeatedly been shown to be an effective tool to do so.

A well-known example is the so-called German tank problem (Ruggles & Brodie, 1947). Suppose that during World War II you were on the planning staff for the invasion of France. Among the questions to be answered is what to bring when we cross the English Channel in our invasion fleet. We only have so much space to carry stuff.

Amid the hazards faced by our troops are tanks. The best protection against these fearsome weapons is bringing our own, but these machines are bulky and heavy. If we bring too few tanks, our troops will be vulnerable; if we bring too many, then we will not have room for bullets, bandages, beans, and other supplies needed to accomplish our mission.

Our intelligence experts draw upon various sources to estimate German monthly tank production, which permits us to estimate the number of tanks available to challenge our invasion force. Some on our staff question the sources and methods used to arrive at these estimates.

Then, a numerate staffer suggests using serial numbers from captured and destroyed tanks to make inferences about monthly tank production. It appears that the serial numbers identify the month of production, so studying samples of known serial numbers could permit forming inferences about the populations of tanks produced each month.

Table 10.1 World War II German Monthly Tank Production

Month	*Intelligence estimate*	*Statistical estimate*	*German records*
June 1940	1,000	169	122
June 1941	1,550	244	271
August 1942	1,550	327	342

Source: Ruggles and Brodie, p. 89.

Table 10.1 compares the two approaches with actual German records. The amazingly accurate predictions put forth from statistical estimates show the potential power of statistical inference.

We may use sample statistics to make inferences about a population's size, average value, standard deviation, or proportion (e.g., share of voters who will vote for a given candidate). For purposes of this book, we'll just stick with ***population mean***. Regardless of what we try to estimate, we need to signal the degree of uncertainty we perceive to be associated with our inference. To this end, we use α, the Greek letter ***alpha***, to communicate our degree of humility.

Assume that our inference could be right or wrong, where the probabilities of these dichotomous outcomes sum to 100%. Let's use the label ***confidence level*** to name the chance that we're right and α to show the likelihood that we're wrong. We could frame these possibilities as follows:

$$100\% = \text{perceived probability of being right} + \text{perceived probability of being wrong}$$

$$100\% = \text{confidence level} + \alpha$$

We may thus define α as 100% – confidence level.

Alpha is a way of showing how worried we are that our predictions will be wrong over repeated samples, also known as ***trials***. To make this clearer, consider the experience of an amateur meteorologist who starts his day with a prediction of whether it will rain. Each morning, our lay weatherman walks into his study to consult his thermometer, hygrometer, wind vane, and barometer to note the current air temperature, humidity, and wind direction plus the overnight change in air pressure. He then uses these four inputs to make a prediction of whether it will rain later in the day.

A comparison of actual results with forecasted values over many trials shows that it does indeed rain about seven days out of every 10 for which a rain prediction is made. Using our formula, we may say that the confidence level for the weatherman's rain forecasts is 70%, so α is 30%.

Should we have 20 future days with the identical temperature, humidity, wind direction, and overnight change in air pressure shown by our instruments today, then we expect that our forecasting model will get it right 14 times and wrong six times. Alpha collapses expected failures for future trials to a single percentage today. Here, there is a 30% chance of being wrong.

We don't expect our forecasting to be perfect because weather patterns are complicated. We know that many factors influence whether it will rain. Our amateur meteorologist could spend time improving his model by testing other variables such as the water temperature of a nearby lake, behavior of the jet stream over Canada, presence of hurricanes in the Gulf of Mexico, and so on. However, meteorology is just one of many hobbies, so he accepts the current level of alpha and chooses to devote time to other pursuits.

This morning he predicts that it will rain. When his wife asks for guidance on how to pack for the day, he responds, *Honey, I think there's a 70% chance of rain; please bring an umbrella.* Using a confidence level signals uncertainty, so his spouse won't be angry if the prediction proves to be wrong.

In this simplistic example, we're working with a dichotomous variable. Let's expand this example to a slightly more complicated setting with discrete variables. Assume we have a big jar full of marbles where each is painted with a number. We're told that the numbers vary around an unknown population average.

We're asked to guess the value of μ based on experiences of drawing a single marble. We do this 20 times before our host reveals that μ equals 50. To summarize, x = the sample mean (the number shown for a single marble drawn in a given trial), n = 20 trials, and μ = 50 (but we don't know this when we draw the marbles). Figure 10.1 presents our results and introduces the idea of ***margin of error***, which communicates an assessment of how far from a predicted value the actual, unobserved number could fall.

Let's start with trial A. We pull a marble with a 50 painted on it. We could infer that μ = 50 but recognize that we're dealing with a single data point. The value of the pulled marble for trial A is shown by the solid black dot. We have one sample with one observation; one's sample cannot get much thinner than this.

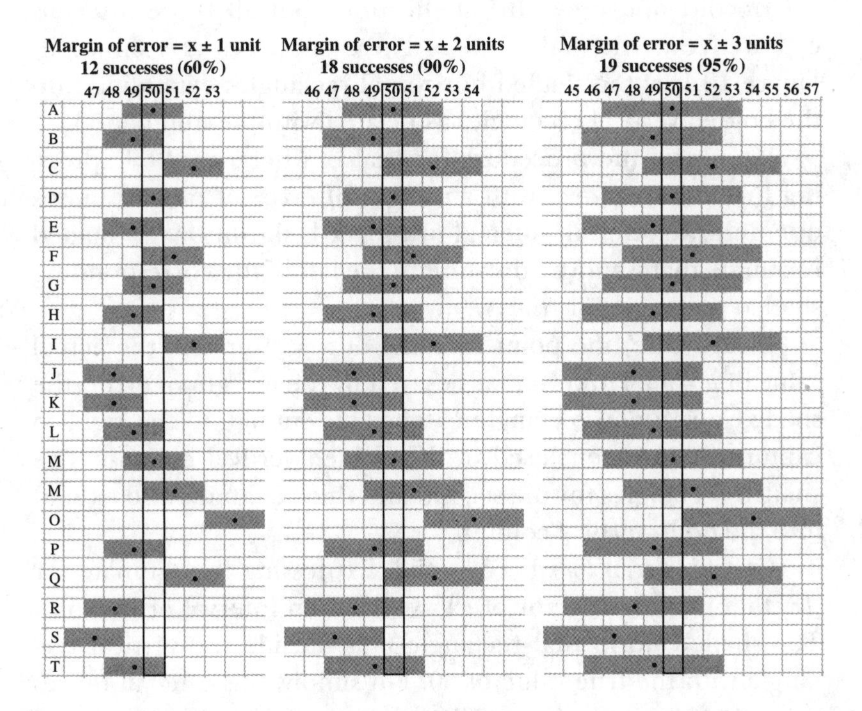

Figure 10.1 Larger Margin of Error Raises Likelihood of Capturing μ

We wish to signal to users of our analysis that there is uncertainty in our inference about the value of μ. One option is to express our inference as a range around the value of the drawn marble. We decide to offer three predictions based on three margins of error. The three columns show ranges of x plus or minus 1, 2, and 3. Thus, for trial A, our three margins of error yield, respectively, intervals of [49, 51], [48, 52], and [47, 53]. The scope of these intervals is shown by the shaded horizontal rectangles.

An interpretation for the third estimate would be, *Boss, our guess is that* $\mu = 50$*, but we have uncertainty associated with this prediction. Let's say that the real value could be as low as 47 but as high as 53.*

The boss may express discomfort with our lack of precision but recognizes that what we provide is better than nothing. Our brief analysis modestly reduces the scope of uncertainty and provides some modicum of information.

We did not know this at the time, but all three intervals captured the true value of μ. These successes are shown in Figure 10.1 by the shaded horizontal rectangles extending into the bolded vertical rectangles associated with the number 50.

We repeat the procedure for trial B, where $x = 49$. Expressing the estimate for μ with a margin of error of plus or minus one unit, we offer an interval of [48, 50], shown by the shaded rectangle that is moved one unit to the left of the corresponding shaded rectangle for trial A.

In this case, the point estimate of $x = 49$ misses the actual value of $\mu = 50$. However, we're saved by the margin of error: the shaded horizontal rectangle extends far enough to the right to capture the area enclosed in the bolded vertical column. The shaded rectangles for the second and third columns also extend into the bold, vertical column.

Consider trial C, where $x = 52$. Expressing the estimate for μ with a margin of error of ± 1, we offer an interval of [51, 53]. The shaded horizontal rectangle falls outside the vertical rectangle with the true value of 50. Put simply, this interval misses the mark. However, the middle column, offering a ± 2 margin of error, provides an interval of [50, 54]. With a larger margin of error, the middle value successfully captures the true value of μ. The third column, with the largest margin of error, also successfully captures μ.

The rest of the table shows results for the remaining 17 trials. Note that in trial S ($x = 47$), only the third column, with the largest of margin of error, has sufficient range to envelop μ. In trial O ($x = 54$), all three intervals fail to capture μ.

Pulling the camera back, we note a positive relationship between an interval's margin of error and its chances of capturing the value of μ. Table 10.2 summarizes our results. The first column shows the width of our intervals; the second, our success rate for each level of width; the third, our confidence level expressed as a percentage out of 20 trials; and the fourth, the value of α, calculated as 1 – confidence level.

Thus, the first row shows that when we offered a simple point estimate, we correctly guessed the value of μ just four out of 20 times, creating a confidence level of 20% and a corresponding α of 80%. As we widened the margin of error, our success rate improved and the value of α went down.

This exercise demonstrates that inference is associated with uncertainty. Using sample data to guess the true value of the population parameter could be problematic because our sample may not be representative of the population so that the sample mean is not close to the population mean. The luck of the draw may cause us to be wrong.

Table 10.2 Increasing Margin of Error Reduces α

Margin of Error	*Number of Successes*	*Confidence Level*	α
±0	4	20%	80%
±1	12	60%	40%
±2	18	90%	10%
±3	19	95%	5%

A commonly accepted way to communicate uncertainty is to show estimated population parameters in terms of a margin of error and an associated confidence level. Here is the idea shown as a formula:

$$\mu = \underbrace{x \pm \text{margin of error}}_{\text{"How wide?"}} \quad @ \quad \underbrace{\text{confidence level}}_{\text{"How certain?"}}$$

The author F. Scott Fitzgerald said that the test of a first-rate intelligence is the ability to consider two opposing ideas and still retain the ability to function. I find this quote comforting as I struggle to think about two different measures of uncertainty. Here we must simultaneously embrace the width of a confidence interval, measured as a margin of error, and then the fraction of trials over which the interval is expected to capture the true value of the thing to be estimated.

To go back to the prior example, results of the middle column could be expressed as follows:

$$\mu = 50 \quad \underbrace{\pm 2 \text{ units}}_{\text{"How wide?"}} \quad @ \quad \underbrace{90\%}_{\text{"How certain?"}}$$

The 90% confidence level means that we're accepting a level of uncertainty (α) of 10%. I admit that I struggle to think this way.

To try to make the idea of combining a margin of error with a confidence level more understandable, let's consider how I could sort through coping with a misplaced mobile phone. After arriving at the parking lot at work, I wish to send my wife a text message so I reach into my pocket, only to find that my phone is not there. A search reveals that the phone is nowhere to be found in my car.

Table 10.3 Increasing the Margin of Error Reduces Alpha

Margin of error *"How wide?"*	*Confidence level* *"How certain?*	*Uncertainty* α
Bedroom closet	60%	40%
Bedroom	75%	25%
House	95%	5%
Ohio	99.9%	0.1%

Sadly, I never learned how to use the device's find-my-phone function. I simply sort through recent memories to try remembering when I last held it. Table 10.3 shows my attempt to use margin of error, confidence level, and alpha to work through this problem.

My most recent memory of holding the phone was just before I changed my clothes in the bedroom closet last night. The phone may still be there. I perceive that I have a 60% confidence level that the phone is within a couple of feet from the center of the closet. This confidence level is equivalent to saying that my level of uncertainty, expressed as alpha, is 40% – a relatively high value, suggesting that I'm not convinced that the phone is indeed there.

I could expand my confidence interval to include our bedroom. It's quite possible that I set the phone down on a dresser after walking out of the closet, extending the margin of error to perhaps 15 feet from the center of the closet. I feel more certain that this wider margin of error captures the location of the phone, so I assign a confidence level of 75%, which lowers α to 25%. Of course, a 1-in-4 chance of being wrong is still meaningful, so I will not say definitively that the phone is within my closet or adjacent bedroom.

Thinking more broadly, I'm pretty sure the phone is somewhere in our house. This margin of error raises the distance from the center of my closet to perhaps 50 feet. A complicating factor is that our bedroom is on the second floor, so, if the phone is truly in our house, it could be on the second floor, main floor, basement, or garage – there are many possibilities. It could take a while to track it down. I assign a 95% level of confidence that the phone is still somewhere in the house, and the resulting α of 5% suggests that there is only a 1-in-20 chance that I'm wrong.

Yet, I'm not positive that I'm right. Among many possibilities is that a family member accidently threw out the phone with the trash, a guest confused my phone with his and took it with him, or a thief absconded with it.

Thinking about these scenarios, I arrive at a margin of error that extends across the state of Ohio, perhaps a radius of 150 miles. I assign a 99.9% probability (with $\alpha = 0.1\%$) that the phone remains within Ohio's borders. I refuse to say with 100% certainty that I may identify the boundary containing the phone. I leave it to your imagination to envision other explanations. The big idea is that raising the margin of error reduces alpha.

So, what is the correct margin of error that we should use to sort through the problem? As with all statistics problems, there is no definitive answer. I liken margin of error to using a butterfly net to capture butterflies. A net that is too small won't give us a very good chance of catching butterflies, but a too-big net is unwieldy and thus useless: there is no way I'm going to spend time searching the state of Ohio for my lost phone.

We have two open questions when using confidence intervals to reduce uncertainty:

1. How do we associate margins of error with varied levels of α?
2. What level of α should we use when responding to questions from our boss?

We'll address these questions, respectively, in Chapters 11 and 12.

Before closing this chapter, I ask that we spend time acknowledging the difficulty of thinking in statistical terms. I have taught the idea of considering margin of error and confidence level simultaneously for years and have yet to complete a semester knowing that all my students got it.

The obvious hypothesis is that I'm a bad teacher, which is something I have not yet falsified. However, I now have enough teaching experience to suggest a second hypothesis: statistical reasoning is phenomenally difficult to do well. I believe that educators have not recognized how hard it is for students to grasp statistical concepts.

If you grew up in the United States, chances are that your teachers nudged promising math students toward an accelerated course of study that provided exposure to calculus before high school graduation. Favored students received instruction in algebra, geometry, trigonometry, limits, differentiation, and integration. The most promising may have also received exposure to multivariate functions, differential equations, and linear algebra before heading off to college.

The rest of us typically received four years of high school mathematics, albeit without the spice. Topics included functions, graphs, data distributions, and introductory probability and statistics. The implicit message was that the smart kids should study calculus while the rest of us should focus on basic data analysis.

I believe that the U.S. secondary education system gets this all wrong. The most useful mathematical training comes from the study of statistical reasoning, not calculus. I spent 30 years at a numerate organization and can name a *single* instance when calculus proved to be useful (we sought to estimate the area under the curve for a time series). Our most significant business decisions, by contrast, came from applying statistical reasoning.

Table 10.4 Calculus Predates Statistics by Two Centuries

1650	*1700*	*1750*	*1800*	*1850*	*1900*	*1950*
Newton		**Euler**	**Green**			
Fundamental Theorem of Calculus		Differential equations	Multivariate analysis			
				Pearson	**Fisher**	**Popper**
				Variance, correlation, and *p*-value	Design of experiments	Falsification

A technology officer recently lamented how his daughter's pre-calculus homework would likely do little to prepare her for jobs in the twenty-first-century workplace. Genuinely useful training, this father believed, would help her learn how to analyze data to improve products, increase sales, or reduce costs. While calculus is necessary for engineering and hard sciences, the study of statistics would better enable students to help organizations accomplish their goals (Markarian, 2018).

Going unnoticed by most math teachers is that understanding statistical reasoning is significantly more difficult than learning calculus. My evidence, presented in Table 10.4, is that humans invented calculus two centuries ahead of statistics. Put simply, it took people a lot longer to learn how to reason statistically than to describe motion and change.

Isaac Newton and his German competitor Gottfried Leibniz seem to have independently created calculus in the mid-seventeenth century. Newton showed how differentiation and integration are inverse operations and is given credit for suggesting the Fundamental Theorem of Calculus. About a century later, Swiss polymath Leonhard Euler showed how to solve many types of differential equations, where one studies small changes and then works backward to define underlying relationships. In the early nineteenth century, English autodidact George Green opened the door for applying calculus to phenomena with multiple variables or functions. By the mid-nineteenth century, calculus had been fully developed.

Because calculus is subject to mathematical proof and its application gave rise to amazing astronomical predictions, this branch of mathematics spurred the development of a point of view labeled the "clockwork universe" (Salsburg, 2001). The thinking went that combining formulas like Newton's inverse-square law with robust celestial measurements would bring the ability to predict all future events.

By the beginning of the twentieth century, scientists came to believe that this was a fool's errand because many phenomena defy prediction, even when observers have access to advanced calculus and lots of data. The rise of quantum mechanics, to take just one example, cast doubts on humans' ability to make predictions or explain things with certainty.

Repeated cycles of data gathering and hypothesis testing have revealed the difficulties with completely understanding the world around us. The best we can do is to think statistically – that is, to extract information from data that reduces but does not eliminate uncertainty.

The word "statistics" entered the English language about 1770 (Hudson, 2000, p. 29), decades after Newton and Leibniz had died. It wasn't until the close of the nineteenth century that English biostatistician Karl Pearson advanced statistical concepts of variance, correlation, and p-value (to be covered in Chapter 14).

After World War I, fellow Englishman Ronald Fisher worked at an agricultural research station and developed modern methods of experimental design. Karl Popper, an Austrian exile who moved to Britain, furthered the idea of falsification, where one accepts that generalizations outside of mathematics may never be proven true but may be shown to be incorrect.

The tools developed by these Britons have helped people reason through messy, uncertain problems like medical research, weather forecasting, investing, advertising, and quality control.

Calculus, for all its theoretical rigor, offers little help as organizations try to find their ways through a messy world. Yet, we steer the smart kids toward the calculus sequence. When you present your findings to important audiences, consider that the education of your listeners was likely grounded in terms of precise limits, derivatives, and integrals and not in the imprecision associated with statistical inference.

Recap of Chapter 10 (Population Mean [μ])

- x is a sample statistic while μ is a population parameter.
- Estimating μ from calculations using x is an example of inferential statistics.
- Estimates of μ require simultaneous consideration of margin of error and confidence level
- α, defined as 100% – confidence level, is a measure of how likely our inference is wrong.
- Increasing the margin of error reduces α.
- Statistics is harder than calculus.

11

CENTRAL LIMIT THEOREM [CLT]

We learned in the previous chapter that statistical inference requires making room in our heads to simultaneously embrace two ideas: the margin of error (how wide?) and the confidence level (how certain?) associated with confidence intervals. Missing from our discussion was a way to associate a given margin of error with a particular confidence level. Chapters 11 and 12 work together to provide a framework to address this problem.

This short chapter relies on pictures to introduce the ***central limit theorem*** (CLT), which says that the distribution of sample means approaches a bell curve as the number of samples increases. In other words, if we take repeated samples from a population, a histogram of sample averages will start to look like a normal distribution.

In this chapter, I use n to signify the number of observations in a sample, x to signify the arithmetic mean of the values in a sample, and the word *trials* to represent the process of taking successive samples from a population.

The CLT is a big deal in the world of numeracy – some say that the CLT is the greatest idea in all of statistics – because it allows us to use the normal distribution's 68/95/99.7 rule to measure the width and certainty of confidence intervals. I emphasize that this chapter is more about visually displaying patterns than about making calculations.

Let's start with an example in Figure 11.1, which shows output of a computer simulation for the CLT. I show static output from a simulation that is better viewed online.[1]

The top panel shows a histogram of a large population. As we learned in Chapter 4, a histogram is a frequency distribution where the *x*-axis shows bins containing different values for the variable being measured and the *y*-axis adds the number of observations in each bin. Bins convert continuous data into discrete buckets that may be easily counted. Here we have a histogram showing bins with integer values ranging from 0 to 32. The population is a normal distribution, shown by a bell-shaped curve with a mean (μ), median, and mode of 16.

Suppose that we wish to estimate the value of μ, but it would be too expensive to measure every member of the population. Instead, we collect random samples from the population, measure the mean of each sample, and plot the sample means on a separate histogram. Each trial involves pulling five members of the population and calculating the arithmetic mean from the five observations.

The second panel down shows results of performing this exercise 100 times. Notice that the mode of the 100 sample means is 15, pretty close to the population mean of 16. Comparing modes of means is a way of taking averages of averages. The shape of the distribution of sample means in the second panel is roughly that of a bell curve.

---

[1]The simulation is available at http://onlinestatbook.com/stat_sim/sampling_dist/ (retrieved 8 December 2020).

Population, $\mu = 16$

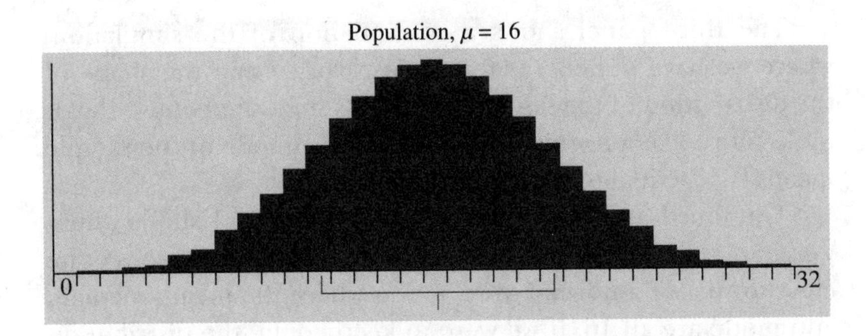

100 trials, $n = 5$, mode of sample means = 15

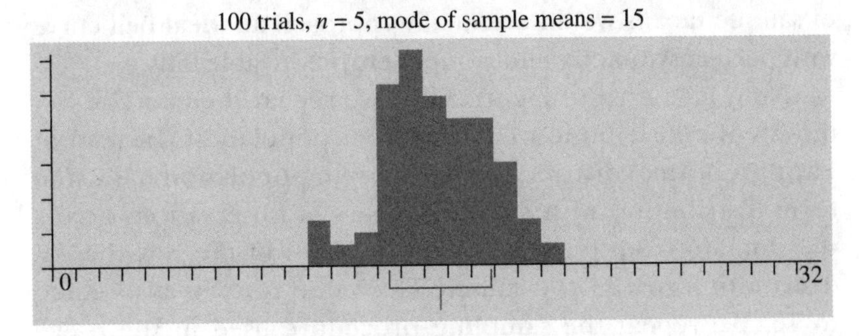

1,000 trials, $n = 5$, mode of sample means = 16

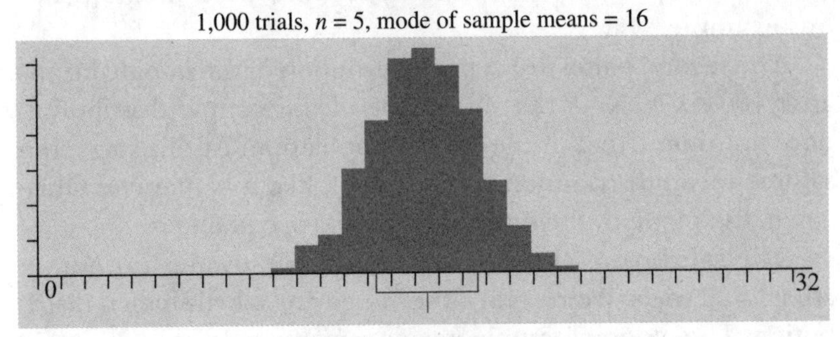

10,000 trials, $n = 5$, mode of sample means = 16

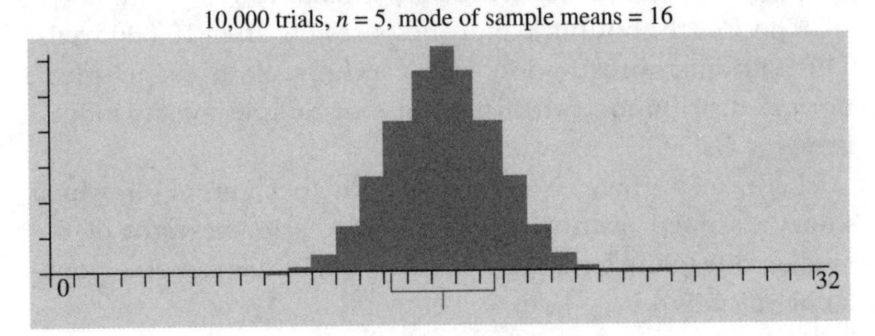

Figure 11.1 CLT Demonstration – Normally Distributed Population

The third panel shows a continuation of the simulation, where we have pulled 1,000 samples. Notice that the shape of the distribution of the sample means has smoothed out so that it looks more like a normal distribution. The mode of the sample means is 16, equal to the population mean.

The fourth panel at the bottom of Figure 11.1 shows where we stand after 10,000 samples. The histogram shape comes quite close to that of a normal distribution where the mean, median, and mode are all 16. If we were to keep going, the distribution of sample means would eventually approach an ideal bell curve with perfect symmetry and completely predictable tails.

Now, let's turn to Figure 11.2. We repeat the exercise but modify the distribution of the parent population. Instead of having a normal distribution, the parent population has a uniform distribution with equal numbers of observations across the domain from 0 through 32, as shown in the box-shaped distribution in the top panel. The value for μ is also equal to 16. We repeat the sampling procedure used in the previous example.

The second panel shows the distribution of the sample means after 100 trials. Note that the shape of the sample distribution does not mimic that of the parent population. At this stage, the sample mean distribution starts to look like a bell curve. Once again, the mode of the sample means is 16, equal to μ.

The third panel down shows the sample mean distribution after 1,000 trials. We're a lot closer to having a bell-shaped distribution. The mode of sample means remains at 16.

The fourth, bottom panel shows results after 10,000 trials. The sampling distribution looks awfully close to an ideal, normal distribution, with the mode of sample means lodged firmly at 16.

Let's keep going. Now, please turn to Figure 11.3, which shows a skewed distribution. The mean is to the right of the median due to the long, right-handed tail. In this case, the population mean is 8.1.

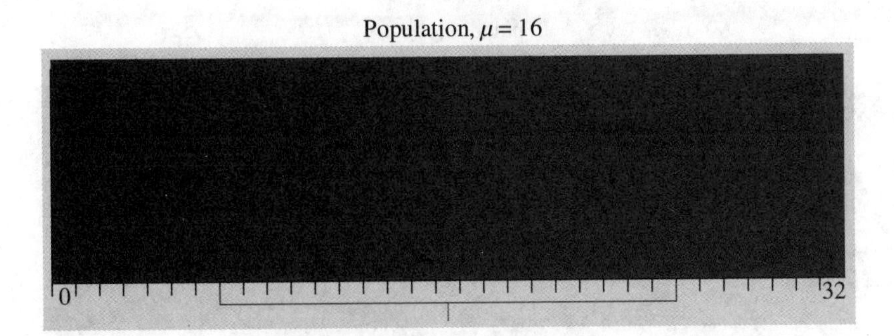

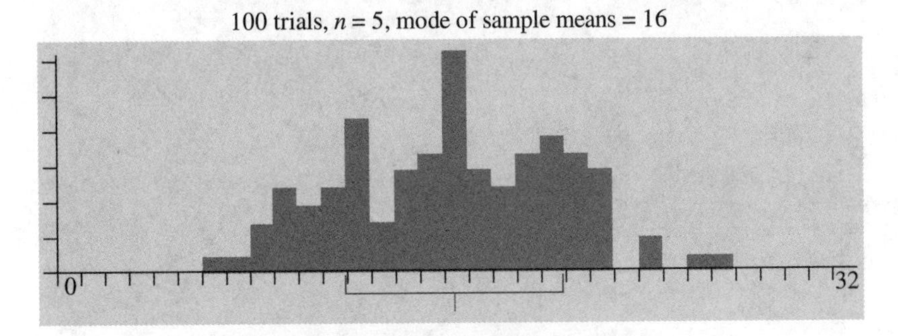

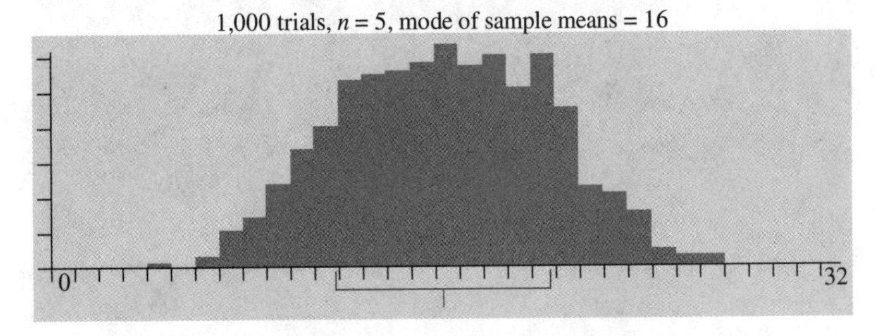

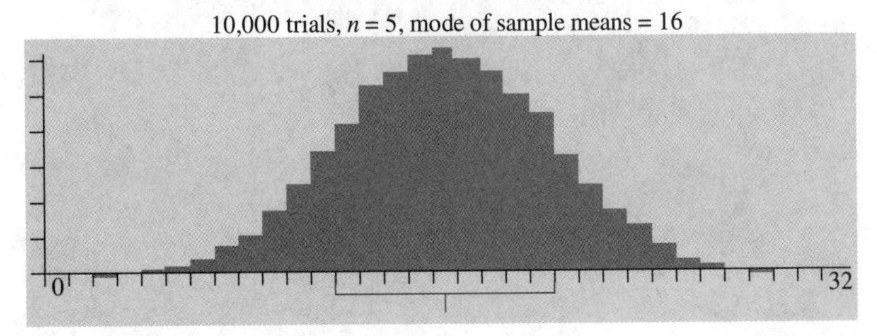

Figure 11.2 CLT Demonstration – Uniformly Distributed Population

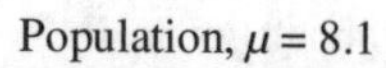

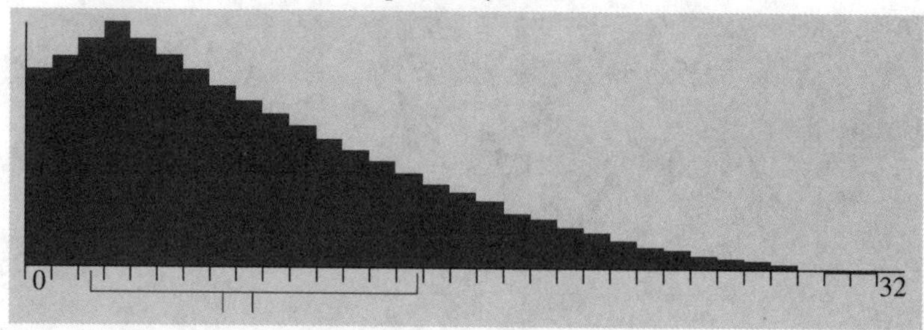

100 trials, n = 5, mode of sample means = 9

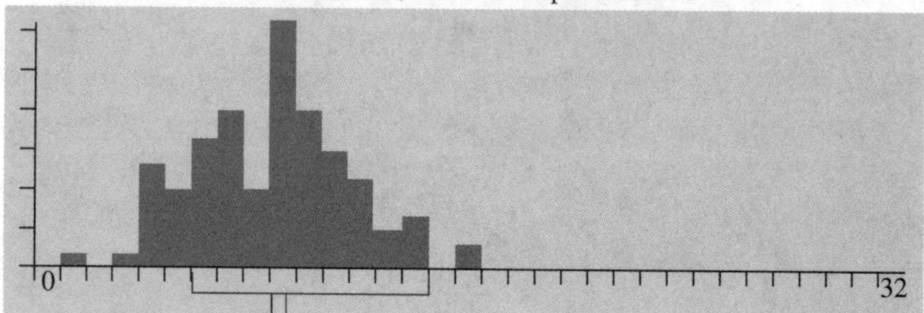

1,000 trials, n = 5, mode of sample means = 8

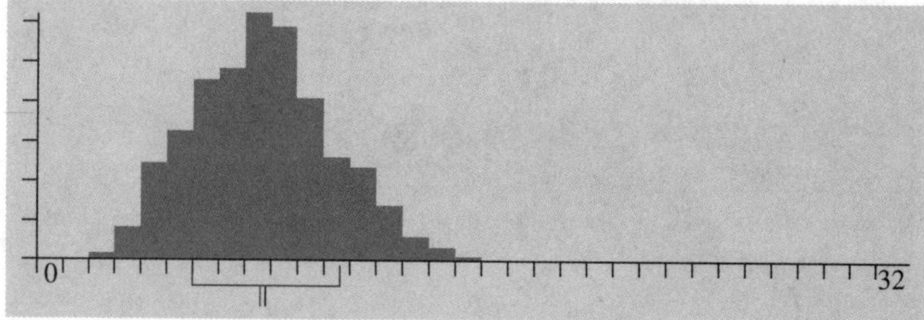

10,000 trials, n = 5, mode of sample means = 8

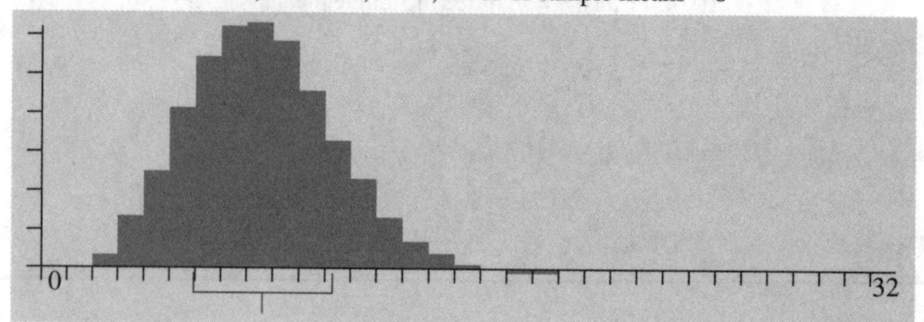

Figure 11.3 CLT Demonstration – Skewed Population Distribution

The second panel shows the distribution of sample means after 100 trials, where the right tail is not nearly as pronounced as it is with the parent populations. The mode of the sample means is 9, a little above the 8.1 value of μ.

The third panel continues with n equal to 1,000 trials. The distribution looks even more like a bell curve. Here, the mode of the means is 8, equal to μ.

The bottom panel shows results of the simulation with n of 10,000 trials. There is a hint of skew, but the shape is even closer to a normal distribution. The mode of the means remains at 8.

Let's do one more. Please look at Figure 11.4, which shows an irregularly shaped population distribution showing little rhyme or reason and a value of μ of 15.9. The next panel shows the distribution of sample means after 100 trials. Note how the mode of sample means is 18. The third panel shows results after 1,000 trials, where the mode drops to 14. Finally, the bottom panel shows results after 10,000 trials, where the mode inches up to 15, not far from the population mean. The distribution of sample means looks pretty darn close to a normal distribution.

Let me now generalize with a pair of findings. First, the means of random samples approaches the population mean as the sample size gets larger. This is known as the ***law of large numbers***, which says that x approaches μ as n increases. This is the idea shown on a Swiss stamp featuring the portrait of Swiss mathematician Jacob Bernoulli (1655–1705), who often receives credit for suggesting the idea. In the graph on the stamp, the sample means start to oscillate around the true population average about a third of the way across the stamp.

An average of means from random samples gets us close to a population mean. Even a few trials helps us reduce uncertainty should we wish to estimate the value of an unobservable population mean. As the number of trials increases, we may take more comfort that x approximates μ.

The second generalization is that the distribution of sample means from randomly drawn samples approximates a bell curve,

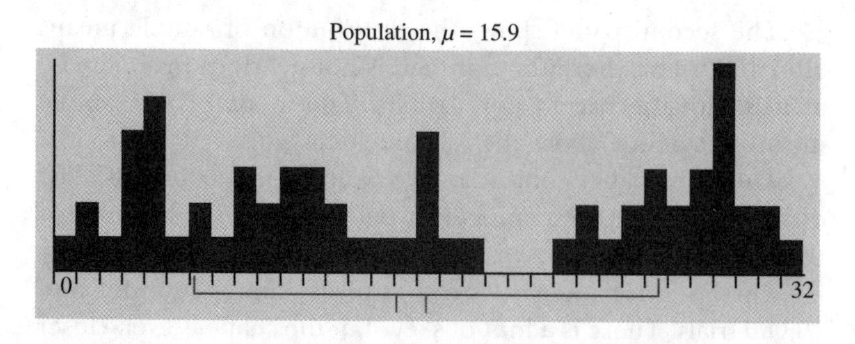

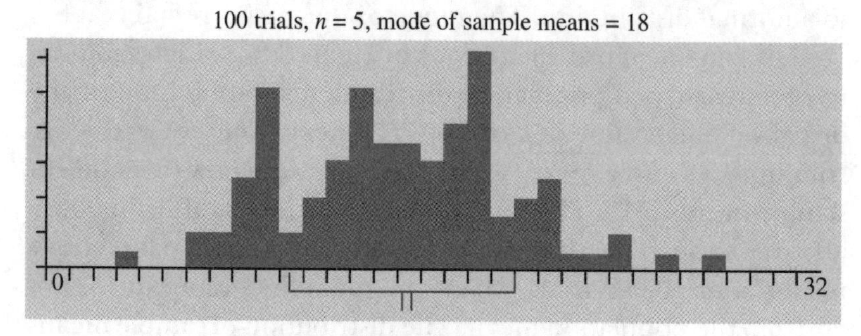

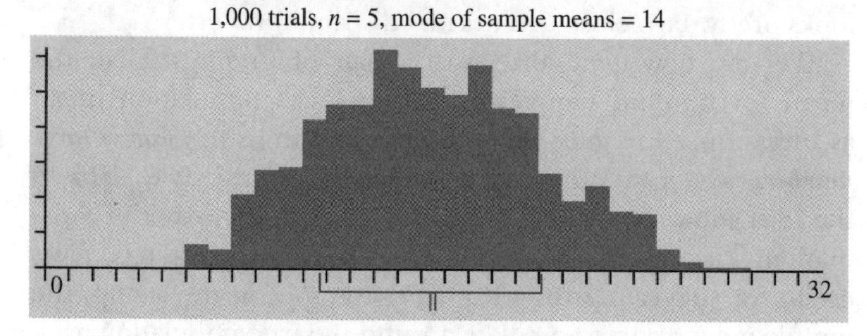

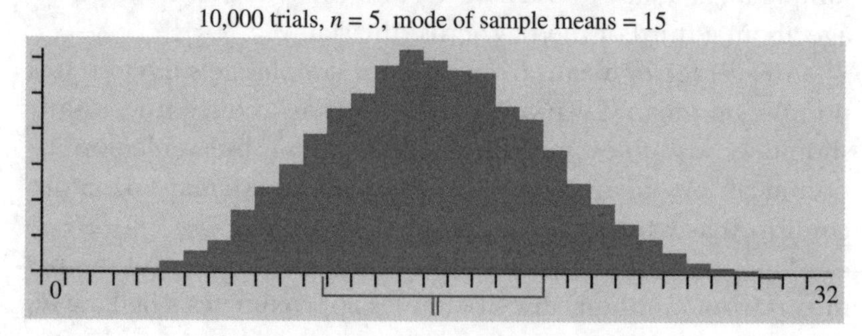

Figure 11.4 CLT Demonstration – Irregular Population Distribution

regardless of the shape of the distribution of the parent population. This is a really cool trick because, as I show in Chapter 12, the bell-shaped curve allows us to use the empirical rule to make inferences about a population mean from sample data.

Let me show one more relationship before closing out the chapter. Figure 11.5 shows what happens when we modify the sample size while holding the number of trials constant.

The top panel shows the parent population, which is again normally distributed with a population mean of 16. The next three panels show distributions of sample means resulting from 100 trials. The difference is the sample size per trial, which were, respectively, 2, 5, and 10. Notice that two things happen as we increase n:

1. The value of x gets closer to μ; and
2. The value of s goes down (shown graphically as narrower spreads).

Put simply, larger sample sizes bring more accurate and tighter estimates of unknown population means. Given the choice between a small or a large random sample, choose the larger sample.

What I want you to take away from this chapter is that if we take a series of random samples, the average of the sample means will cluster around the population mean in a bell-shaped distribution. The conceptual leap is viewing one sample of, say, 30 observations as 30 trials of one observation per sample. The means of the 30 sampled, individual observations, if drawn randomly, will form the early stages of a normal distribution distributed around the unobserved population mean.

A larger sample size bestows two benefits. If we increase the sample to 50 observations, the sample mean will likely fall closer to the true population mean. Further, the distribution of individual observations will cluster more tightly around the average of the sample means. In other words, a larger sample will suggest an x that is closer to μ with a smaller s.

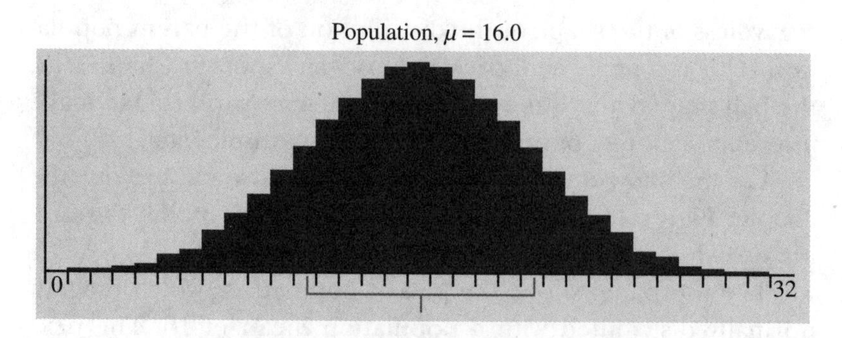

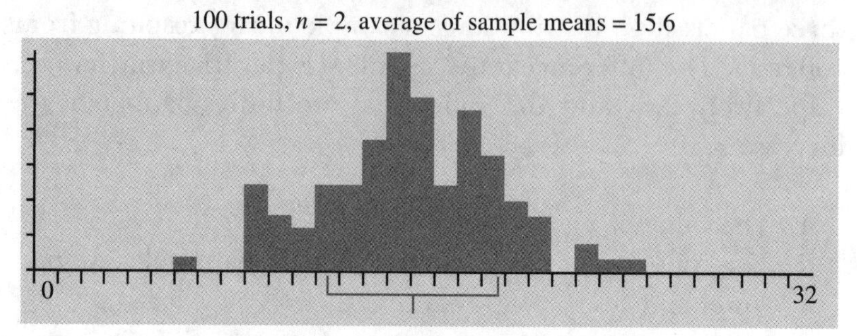

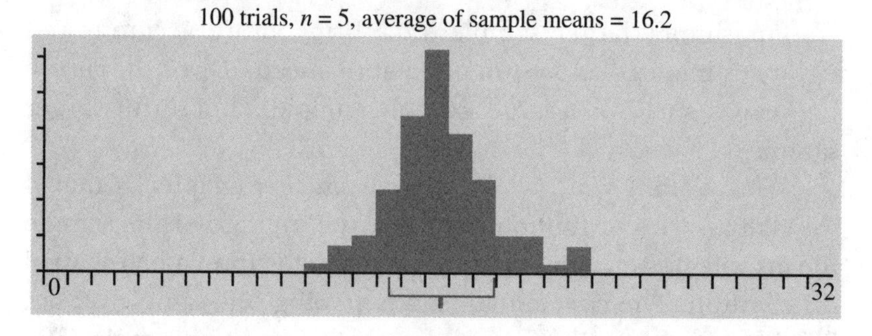

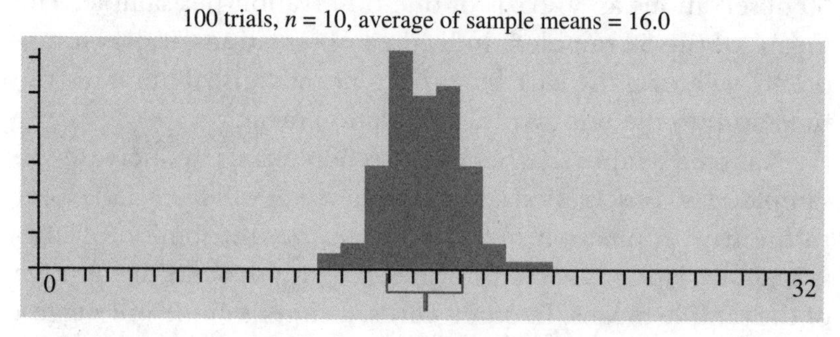

Figure 11.5 CLT Demonstration – Varied Sample Sizes per Trial

I offer two big caveats for using the central limit theorem. First, the concept rests on using random samples, which I've argued is an idea rather than something you'll ever do.

Second, inferences apply to the population at the point in time that the sample is made. If a population undergoes a change – say, investors look at the world differently after the investment bank Lehman Brothers filed for bankruptcy – then inferences made from sampling completed before the change may be misleading. A sampling of investor sentiment performed in August 2008 would likely be of little use to make predictions about investor behavior in October 2008.

Recap of Chapter 11 (Central Limit Theorem [CLT])

- The CLT is based on the elusive idea of random samples.
- Accumulated sample means from random samples approximate a normal distribution.
- A larger sample brings x closer to μ with a smaller s.
- All bets are off if the population changes after samples have been taken.

12

STANDARD ERROR [$s/\sqrt{n}$]

In Chapter 10, we introduced the idea of inferential statistics, where we seek to estimate the size of an unknown population parameter based on observations from samples. In Chapter 11, we introduced the central limit theorem (CLT), which states that distributions of sample means from random samples gravitate toward normal distributions, regardless of the shape of parent populations. In this chapter, we tie things together to show one way to quantify uncertainty when making inferences.

Our main attraction is the idea of ***standard error***, defined as $s/\sqrt{n}$, that measures distances of x from μ. Standard error is the basis for assessing uncertainty associated with inferences. We use standard error to help answer the question of how wide a confidence interval should be. As shown shortly, an inference's margin of error is the product of the number of required standard errors multiplied by the size of the standard error.

Figure 12.1 presents an illustration showing the logic. The top oval represents the population of whatever interests us. The population is the entire constellation of data points at a point in time. The *group* typically is a collective noun – a collection of people, places, or things – that we seek to better understand through quantification.

It is often too expensive or time-consuming to measure every item in a population, so we settle for measurements of a sample of the people, places, or things taken from the population. We use measurements of samples (sample statistics) to make inferences about values of things that interest us in the population (population parameters).

In this case, the population parameter to be estimated is the average value for a trait for members of the population. Let's assume that the population parameter to be estimated is height, shown as μ. We are able to measure the heights of members of a sample and then calculate three familiar sample statistics: n, the number of observations in the sample; x, the arithmetic mean of the heights of all observations; and s, the sample standard deviation of the observed heights. Put simply, using our old friends n, x, and s helps us make an inference about μ.

Suppose we seek to join a social fraternity on a college campus. The pledge master, the fraternity's officer responsible for screening candidates, designs tasks to identify those who have the right stuff for admission. We've chosen to rush the nerd fraternity, and the pledge master has gathered us at the frat house on a crisp fall morning to present our first assignment. We're handed rulers and given 30 minutes to tell him the average height of all blades of grass on the lawn enveloping the house.

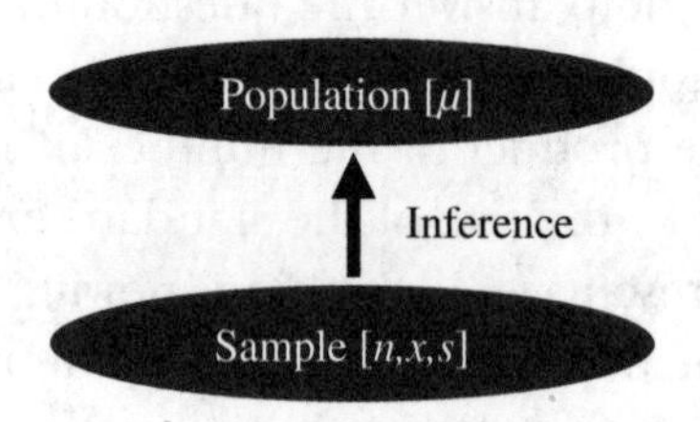

Figure 12.1 Inferences Derive Population Parameters from Sample Statistics

There is no feasible way (an approximate translation of what one of the pledges mutters) that we can measure every blade of grass in the allotted time. Some whiny recruits head to the nearby beer keg in efforts to bypass the pledge master and schmooze the fraternity president.

We're better than this, say *No problem, boss,* and draw on skills learned in previous chapters. Before proceeding, we ask for a pencil, pad of paper, and statistics calculator. Our host delivers all three blindingly fast. *Wow,* we think, *perhaps we could be happy here.*

We gather our wits by reflecting on key lessons covered so far:

- In Chapter 3, we learned that it may be possible to make inferences about an unobservable population from an observable sample so long as we avoid the twin problems of the sample being too small and not representative of the population. We remember that the rule of thumb for a decent sample size is 30.
- In Chapter 4 we learned that an arithmetic mean (x) may serve as an effective summary measure for a continuous, ratio variable. Lengths of grass blades fit this classification.
- In Chapter 5, we learned that standard deviation allows us to measure variance around a mean. This will be helpful because we can already tell that the grass blades have varying lengths.
- In Chapter 6, we learned about the empirical rule associated with normal distributions, where about 68%, 95%, and 99.7% of all observations fall, respectively, within one, two, and three standard deviations from the mean.
- In Chapter 10, we learned that estimates of μ should be expressed in terms of confidence intervals that address the two questions of "how wide?" and "how certain?"
- In Chapter 11, we learned that sample means from multiple random samples are distributed normally around the population mean, regardless of the distribution shape of the population.

- At the beginning of Chapter 12, we learned that standard error ($s/\sqrt{n}$) is an adjusted standard deviation used to measure distances of x from μ. A margin of error is the product of the number of required standard errors multiplied by the size of the standard error.

This is going to be fun.

We select 30 blades of grass from around the yard to approximate a random sample. We include a couple of blades from bald spots in unshaded areas where the sun has baked the lawn plus a couple more from lush areas that seem to have been fertilized by Max, the frat's Old English Sheepdog.

We record the lengths of the blades and then use the calculator to compute x and s. Our results are as follows, rounded to a single decimal place:

$n = 30$ blades of grass
$x = 3.8$ cm
$s = 1.2$ cm

We note that inference requires that we convert a standard deviation into a standard error (SE) to measure distances of x from μ. We calculate standard error as follows:

$$\text{SE} = s / \sqrt{n} = 1.2 / \sqrt{30} = 0.2 \text{ cm}$$

We decide to use a 95% confidence level (or $\alpha = 5\%$) to frame our answer. We know that in a normal distribution, a 95% confidence level is associated with about two standard deviations.

Pulling it all together, we jot down the following on our pad of paper:

$$\mu = x \underbrace{\pm \text{ margin of error}}_{\text{“How wide?”}} \; @ \; \underbrace{\text{confidence level}}_{\text{“How certain?”}}$$

Plugging values into this framework using two standard errors, we arrive at:

$$\mu \approx 3.8 \quad \underbrace{\pm 2(0.2)}_{\text{“How wide?”}} \quad @ \quad \underbrace{95\%}_{\text{“How certain?”}}$$

The confidence interval is [3.4, 4.2]. We raise our hand to show we're ready. The pledge master comes over to interrogate us.

So, what do you got, Pledge?

Sir, my estimate is that the population mean is 3.8 centimeters with a margin of error of plus or minus 0.4 centimeters at a 95% level of confidence.

Pledge, how did you get there?

Well, Boss, I used a judgmental sample of blades from around the lawn to mitigate the risk of sample bias. I chose a sample size of 30 to reduce the problem of having a thin sample. I then calculated the sample mean, standard deviation, and standard error. Deciding upon a 95% confidence level, I used the central limit theorem to arrive at a margin of error based on two standard errors. The rest was just plug and chug.

If we didn't know better, we could swear that the pledge master just revealed a small smirk of admiration. Some of the frat brothers, sensing something interesting happening, saunter from the keg to watch the show.

Pledge, I'm not satisfied. I need more certainty. I want to know your answer to a 99% level of confidence.

Well, Sir, I can give you an approximate answer quickly. If we use the central limit theorem to assume that the distribution of sample means is bell-shaped and draw on the empirical rule, which associates three standard deviations with a 99.7% level of confidence, then three standard errors give a distance of about 3 times 0.2 cm or roughly 0.6 cm. Adding and subtracting this distance from the sample mean results in a confidence interval that extends from 3.2 to 4.4 cm.

Some of the frat brothers show approval with hoots and hollers. Everyone else at the rush event comes over to see the excitement. The pledge master, unwilling to show weakness, presses on.

Nice try, Pledge, but your margin of error of plus or minus 0.6 cm is too wide. I need a narrower interval.

We don't panic. We note that standard error is calculated by dividing the sample standard deviation by the square root of the sample size. We also note from Chapter 10 that values for confidence level and uncertainty (or α) are complements; that is, they add up to 1.

Boss, that won't be a problem. I'll give you two choices. First, if you give me more time, I'll increase the sample size beyond 30. A higher n will shrink the standard error because the ratio will be calculated with a larger denominator. Alternatively, we can increase alpha. For example, if you are willing to live with a level of uncertainty of 32%, our resulting confidence level of 68% (or 1 minus 32%) requires that we use just one standard error in our margin of error. The confidence interval would be 3.8 cm plus or minus 0.2 cm or the interval from 3.6 to 4.0.

The crowd goes crazy, with frat brothers dropping to their knees, raising and lowering their heads and arms in unison while chanting *We are not worthy!* Okay, I'm getting carried away.

This imaginary dialogue shows how piecing together some simple concepts and grade-school math puts anybody on the path to numeracy. The exchange reveals a conversation between two numerate people who are working together to extract information from raw data in an iterative process that seeks to reduce uncertainty.

What we have done is to make an inference about an unknown population parameter from known sample statistics. Statistical inference may be used to estimate values other than a population mean. We may, for example, seek to estimate a population proportion (e.g., how many voters are likely to vote for a given political candidate if the election were held today) or a population size (e.g., how many people live in the United States as of a census date).

You will need a propeller head to work through complexities of sample design and required math, but the idea of statistical inference is the same, regardless of what we're trying to estimate. The point of this chapter is to convey that we may reduce uncertainty with some simple calculations. If our educated guess proves to be surprising – that is, appears to yield information – then we should bring in experts to attack the problem with rigor.

Let's try another example to illustrate assumptions embedded in our reasoning. Assume we work in a manufacturing plant that makes a small product. We pull a sample (n) of 64 pieces to measure the weight of each item. We calculate a sample mean (x) of 10 grams and a sample standard deviation (s) of 4 grams.

Already we know that there is some variance in our manufacturing process because the coefficient of variation ($s/x = 4/10 = 40\%$) puts us in the yellow-light range for the traffic light metaphor given in Figure 5.2. It is risky to simply use x to reach a conclusion about the population mean.

Well, how dangerous is it to generalize from x? Our new framework blends the ideas of a confidence interval and level of certainty. Unless otherwise directed, we start with the convention of setting α at 5% (i.e., 95% confidence level) and perform the plugging and chugging:

$$
\begin{aligned}
\mu &\approx x \pm 2\left(s/\sqrt{n}\right) @\, 95\% \text{ confidence level} \\
&\approx 10 \pm 2\left(4/\sqrt{64}\right) \\
&\approx 10 \pm 2(0.5) \\
&\approx [9,11]
\end{aligned}
$$

Based on our sample, we infer that the unobserved population mean is between 9 and 11 grams at a 95% confidence level. The margin of error (how wide?) comes from multiplying two times the standard error, and the level of confidence (how certain?) comes from assuming that two standard errors captures

95% of the observations in a bell curve. If we were to take a sample of 64 units 20 times, we expect that our confidence interval would capture the true population parameter in 19 (95%) of those trials.

If we sought a greater level of confidence, we could use three standard errors to arrive at a margin of error of ± 1.5 grams, which would be associated with a confidence level of about 99.7%. If we sought a narrower confidence level, we could use a single standard error to give a margin of error of ±0.5 grams with a confidence level of about 68%. Increasing (decreasing) the margin of error increases (decreases) the likelihood that an unknown population parameter is captured by the confidence interval.

As mentioned earlier, a metaphor for margin of error is the size of a butterfly net. Capturing a butterfly is not easy because these maneuverable insects often change the speed and direction of flight. One cannot know a butterfly's location at a future point in time with certainty. Using a net allows a lepidopterist (not only do I have a thesaurus, but I'm not afraid to use it) to capture a butterfly based on an imperfect guess about its future location.

The question is, how big a net do you want? A wider net provides a greater prospect of capturing a butterfly in any given swipe. However, at some point, a net's size makes the tool unwieldy and useless. Too-small and too-large nets are not helpful.

The goal of statistics is to reduce uncertainty. Good inferences have reasonable margins of error. A too-small margin of error has little likelihood of capturing a population parameter. A too-large margin of error offers little help in making decisions.

When working with a new client, I argue that we should start with an alpha of 5%, which gives us a 95% level of confidence and – assuming we have a decently sized sample and the central limit theorem holds – a margin of error of two standard errors. We would then make adjustments should we be worried about consequences of false positives or negatives, topics covered later in this book.

Numerate people don't just run numbers; they ask good questions after an analysis has been performed. When handed results of an inference, please ask how the preparer obtained assurance that the sample is representative of the population. Consideration of sample bias weighs as much as concerns about sample size.

The CLT is based on random samples. As discussed earlier, random sampling is a construct, not something found in nature. There was no way that our fraternity pledge could draw a random sample of blades of grass. He was under severe time constraints, and grass blades on the lawn were not numbered. You will face similar problems in your career.

I doubt that there has ever been a perfectly random sample in the history of statistical inference. Numerate people should assume that samples are biased, consider how possible biases could influence statistical findings, and then qualify conclusions with qualitative commentary. A few minutes of discussion and a resulting one-sentence disclaimer could have saved the managing editor of the *Literary Digest* substantial embarrassment discussing results of his organization's 1936 presidential poll.

Perhaps the most significant unrecognized source of bias is failure to consider that the world changes from the time a sample is collected and an inference is presented. Going back to the pledge example, the grass continues to grow during the fall semester. An inference made three days after measurements were taken would be misleading: a three-day-old sample is not representative of today's population. Please use care when drawing inferences from time series data because dated data points may become stale.

The next topic of discussion should cover sample size. We learned in the previous chapter that larger samples bring tighter, more accurate predictions of population parameters. We all would prefer the comfort associated with larger sample sizes. However – and this is something rarely discussed in statistics textbooks – sampling is costly.

Remember that standard error comes from dividing the sample standard deviation by the square root of the sample size ($s/\sqrt{n}$). If we wished to halve a standard error, we would have to *quadruple* the sample size.

Going back to the earlier manufacturing example, suppose that our head of marketing is not comfortable with the one-gram width of the margin of error. She wants to know the average weight of our product to within half a gram before making any promises to her customers. Let's suppose that she is also unwilling to modify assumptions about uncertainty (α = 5%, level of confidence = $1 - \alpha$ = 95% and number of standard errors required for 95% confidence = 2).

If we double our sample size from 64 observations to 128 and assume that the sample mean (x = 10 grams) and standard deviation (s = 4) stay about the same, then our estimate for standard error drops from 0.50 ($s/\sqrt{n} = 4/\sqrt{64} = 0.50$) to 0.35 ($4/\sqrt{128}$). To halve the original standard error, we would have to increase the sample size to 256 units ($0.25 = 4/\sqrt{256}$).

Our head of manufacturing complains that raising our sample size by a factor of four would take too much time, putting production schedules at risk. The president must step in to decide what, if anything, we should do. Statistical inference does not give an answer; instead, this analysis elicits information that sparks more informed discussions. Numerate people layer math with judgment.

Some data come to us from outside of our organization, where we have no ability to influence sample size. This can be a problem because we've learned that observation counts below 30 bring thin samples, which impairs our ability to draw inferences. The additional uncertainty from a very small sample prevents us from simply multiplying standard error by 2 to arrive at a margin of error associated with a 95% confidence level.

A clever English statistician named William Gossett (1876–1937) provided a solution to the small sample problem. He thought about inference while holding one of the coolest jobs in the world, chief brewer of Guinness in Dublin. I grew up near Milwaukee, Wisconsin, so beer drinking has been a hobby since I was an underage teenager. It is difficult to develop a manufacturing process that consistently yields a desired taste due to the many variables that influence brewing.

Gossett oversaw countless experiments and developed a framework for drawing meaningful conclusions from small samples. He sought to publish findings in academic journals, but management was worried about competitors using these techniques to weaken Guinness's competitive position. Gossett was allowed to publish as long as he did not mention beer, Guinness, or his own name (Ziliak, 2019).

Today, you can find Gossett's legacy in "*t* distribution" tables, which appear online and in every introductory statistics book. I caution that they instill fear because they display exhausting numbers of rows, columns, and significant digits. My experience is that they are not worth the bother. Table 12.1 offers a liberal arts introduction to *t*-tables to emphasize the big idea and avoid getting lost in the particulars.

The table shows how many standard errors are required to develop a confidence interval from a random sample. There are two inputs and one output. The inputs are the number of observations in the sample (shown under the column with the heading "*n*") and the desired confidence level for the inference (shown in the columns below the heading "Confidence Level and [α]"). Propeller heads reading these pages will complain that I replaced the traditional column heading "Degrees of Freedom" with *n*; to keep this simple, I assumed that our analysis would chew up just one degree of freedom. I also kept things simple by limiting discussion to a "two-tailed" test, something we touch on later in the book.

To give an example, if we had a sample size of 30 and sought to make an inference about a population mean with a 95% confidence level ($\alpha = 1 - 95\% = 5\%$), then we would use 2.05 standard errors:

Table 12.1 Number of Standard Errors Required for Varied Sample Sizes

	Confidence Level and [α]				
n	*50%* *[50%]*	*68%* *[32%]*	*90%* *[10%]*	*95%* *[5%]*	*99.7%* *[0.3%]*
2	1.00	1.82	6.31	12.71	212.21
3	0.82	1.31	2.92	4.30	18.22
4	0.76	1.19	2.35	3.18	8.89
5	0.74	1.13	2.13	2.78	6.43
10	0.70	1.05	1.83	2.26	4.02
15	0.69	1.03	1.76	2.14	3.58
20	0.69	1.02	1.73	2.09	3.40
25	0.68	1.02	1.71	2.06	3.30
30	0.68	1.01	1.70	**2.05**	3.24
40	0.68	1.01	1.68	**2.02**	3.17
50	0.68	1.00	1.68	**2.01**	3.12
60	0.68	1.00	1.67	**2.00**	3.10
70	0.68	1.00	1.67	**1.99**	3.08
80	0.68	1.00	1.66	**1.99**	3.06
90	0.68	1.00	1.66	**1.99**	3.05
100	0.68	1.00	1.66	**1.98**	3.04
∞	0.67	0.99	1.64	**1.96**	2.97

$$\mu \approx x \pm (2.05)(\text{standard error})\ @\ 95\%\ \text{confidence level}$$

If we increase the sample size above 30 observations and thus travel down this column, the number of required standard errors drops somewhat. Fewer standard errors give rise to narrower confidence intervals required to capture the unknown population parameter at the desired level of confidence. A large sample means that we may use a narrower butterfly net. Think back to Figure 11.5, where expansion of sample size brought narrower distributions of sample means. If we had a huge sample, denoted by the symbol for infinity (∞), then our confidence interval would shrink to ±1.96 standard errors.

If we are forced to work with a smaller sample, then we move up the column. So, if all we had was 15 observations and still sought a confidence level of 95%, our margin of error would expand to ±2.14 standard errors. If we had just a few observations, then our confidence interval would become quite wide.

Holding sample size constant at $n = 30$, look what happens when we raise and then lower our desired level of confidence. If we wished to make an inference at a 99.7% level of confidence, then the number of standard errors would expand to 3.24. This means we're required to use a wider net if we wish to capture a population parameter with this greater level of certainty. If we're willing to live with more uncertainty, we may move to the left of the table. Should we be satisfied with a 90% confidence level ($\alpha = 10\%$), then the width of the confidence interval shrinks to 1.70 standard errors.

I know, this is a lot of numbers. Figure 12.2 distills these relationships into three ideas. First, notice that the number of required standard errors to achieve a given confidence level decreases as the sample size, n, increases from Low to High: larger samples permit more confident inferences. Second, notice that the number of required standard errors decreases as our acceptable level of uncertainty, α, goes from Low to High.

Finally, as sample size approaches 30, we can simply remember that we need about two standard errors to reach a 95% level confidence for inferences drawn from random samples. This is the reason I bolded the entries in the 95% column where n equals 30 or higher. If you must work with a sample of fewer than 30 observations, just qualify your answer that the margin of error would be bigger than two standard errors. If your boss wants a higher level of certainty, you can just say that plus or minus three standard errors would get us to a confidence level percentage in the high nineties.

What I don't want you to do is to use a t-table to develop a confidence interval with lots of decimal places. Doing so offers a false sense of precision because we have almost certainly violated

α

High ← Low

Confidence Level and [α]

n: Low → High

n	50% [50%]	68% [32%]	90% [10%]	95% [5%]	99.7% [0.3%]
2	1.00	1.82	6.31	12.71	212.21
3	0.82	1.31	2.92	4.30	18.22
4	0.76	1.19	2.35	3.18	8.89
5	0.74	1.13	2.13	2.78	6.43
10	0.70	1.05	1.83	2.26	4.02
15	0.69	1.03	1.76	2.14	3.58
20	0.69	1.02	1.73	2.09	3.40
25	0.68	1.02	1.71	2.06	3.30
30	0.68	1.01	1.70	**2.05**	3.24
40	0.68	1.01	1.68	**2.02**	3.17
50	0.68	1.00	1.68	**2.01**	3.12
60	0.68	1.00	1.67	**2.00**	3.10
70	0.68	1.00	1.67	**1.99**	3.08
80	0.68	1.00	1.66	**1.99**	3.06
90	0.68	1.00	1.66	**1.99**	3.05
100	0.68	1.00	1.66	**1.98**	3.04
∞	0.67	0.99	1.64	**1.96**	2.97

Figure 12.2 Larger α and n Reduce Required Number of Standard Errors

the assumptions of using a random sample and working with a stable population that did not change over the period from sample collection to making a decision from our inference.

The ancient Greek philosopher Heraclitus (c. 500 BCE) is alleged to have said that no man steps into the same river twice. His point was that events unfold in ever-changing ways. We cannot expect that what happened yesterday will recur in exactly the same way tomorrow. We use inference as a crude way to quantify predictions of the future from analysis of the past.

I close this chapter with a real example of using inference to make a family decision. Our home lies smack in the middle of Cleveland's snowbelt. Foreseeable but unpredictable snowstorms require periodic snowplowing to clear our driveway over the winter.

Our snowplow contractor offered us a choice of paying a flat fee of $418 for the winter season or a variable charge of $19 per "push" every time there is a snowfall of at least two inches in our community. Breakeven volume is 22 pushes per season. If the next winter brings 23 or more sizable snowfalls, then we're better off paying the flat fee.

My wife keeps good records and notes that over the previous four winters, we had, respectively, 15, 15, 8, and 12 pushes. She asks, *So, Professor, what should we do?* Great question, because we're working with a very thin sample of four observations.

I determine that $n = 4$, $x = 12.5$, $s = 3.3$, and $\text{SE} = 3.3/\sqrt{4} = 1.7$. I deliberately round results to one decimal place to remind myself that this analysis is fraught with problems. I decide to use a 95% confidence level to frame this analysis, consult Table 12.1, and note that four observations requires that we use about 3.2 standard errors to develop a confidence interval. I make the following inference about the "true" mean of pushes per winter in our community:

$$\mu \approx 12.5 \pm (3.2)(1.7) \text{ @ } 95\% \text{ confidence level}$$

The confidence interval extends from 7.1 to 17.9. Since the number of pushes is a discrete variable, I round further to whole numbers to arrive at an interval from 7 to 18 pushes. The breakeven number of pushes is outside the confidence interval.

I then ask what could go wrong with this analysis. Two concerns that come to mind are that global warming trends may make the future different from the past and that paying per push may influence the contractor to plow more frequently than specified by the contract. I have no easy way to quantify these considerations. It's time to make a decision based on incomplete information.

Honey, let's pay by the push.

It turns out that we had 13 pushes that winter. My wife was not impressed.

Recap of Chapter 12 (Standard Error [$s/\sqrt{n}$])

- Standard error (SE), calculated as $s/\sqrt{n}$, assesses the distance that x lies from μ.
- Margin of error ("how wide?") = (SE)(number of required SEs).
- The number of required SEs turns on n and a desired level of α.
- Narrowing the margin of error requires increasing n or α.
- Just remember that two SEs is associated with $\alpha \approx 5\%$ if $n > 30$.

13

NULL HYPOTHESIS [H_0]

A psychologist offered these statements to predict how you view yourself:

While you have some personality weaknesses, you are generally able to compensate for them. Disciplined and self-controlled outside, you tend to be worrisome and insecure inside. At times you have serious doubts as to whether you have made the right decision or done the right thing.

You prefer a certain amount of change and variety and become dissatisfied when hemmed in by restrictions and limitations. You pride yourself as an independent thinker and do not accept others' statements without satisfactory proof.

You have found it unwise to be too frank in revealing yourself to others. At times you are extroverted, affable, and sociable, while at other times you are introverted, wary, reserved. Some of your aspirations tend to be pretty unrealistic. Security is one of your major goals in life. (Forer, 1949)

These statements likely resonate with you. Does their use reflect good science? Almost every academic I know would say *No* because these sentences are so vague that they are difficult to refute.

This chapter introduces the idea of the ***null hypothesis***, signified by H_0, which expresses predictions in a way that they may be disproven. In a word, this chapter is about ***falsification***, developed by philosopher Karl Popper, whom we met in Chapter 9. He said that we could see an unlimited number of white swans but still fail to prove definitively that all swans are white. However, if we find just a single black swan, we have disproven the idea.

It's difficult to prove something but comparatively easy to show that it's wrong. Ruling out explanations that are wrong allows us to channel more attention to a smaller number of surviving relationships that may prove to be useful.

Before going into detail, let's define six more words to be used in the balance of this book:

1. ***Theory:*** A framework that combines ideas to explain the past and/or predict the future.
2. ***Construct:*** An idea that is difficult to define and measure.
3. ***Proposition:*** A suggested relationship between constructs used in a theory.
4. ***Model:*** A simplified theory that may be tested.
5. ***Variable:*** An attempt to define and measure a construct.
6. ***Hypothesis:*** A proposed relationship between two or more variables.

A theory is a mental framework that allows people to make sense out of the world around us. I have never driven to El Paso, Texas, but my own theory of how U.S. roads are organized gives me confidence that I could accomplish this task should I choose to do so. The use of theory empowers surgeons to perform untried operations, attorneys to litigate cases with novel fact patterns, and chefs to prepare meals from substitute ingredients when faced with unexpected shortages.

The word *theory* gets a bad rap outside of university campuses. A surefire way to insult a professional is to say that a proposal may work well in theory but not in practice. Such criticism misses the point that theories serve a critical role in calming the human mind by offering wanted structure to a messy world. Even theories that are wrong are helpful.

Consider the story of Hungarian soldiers who ventured out on maneuvers in the Alps and became enveloped in a blinding snowstorm (Weick, 1995, pp. 54–55). Two days of fatigue and disorientation brought the loss of hope for a safe return home. Then one person found a map. The discovery brought a calming influence. The squad formed a plan to pitch camp for the night and then use the map to find their way home the next morning.

The group successfully returned to their unit and safety. An officer asked to see the document and noticed that it was a map of the Pyrenees, not the Alps. If used literally, the diagram was useless. However, the map offered a sense of order that inspired the platoon to solve their problems. An imperfect theory saved their lives. You, in your career, use theories that you know are not completely true. The fact that they are imperfect does not mean that they should be discarded.

The purpose of a theory is to help us explain the past and/or predict the future. Figure 13.1 shows a classification system.

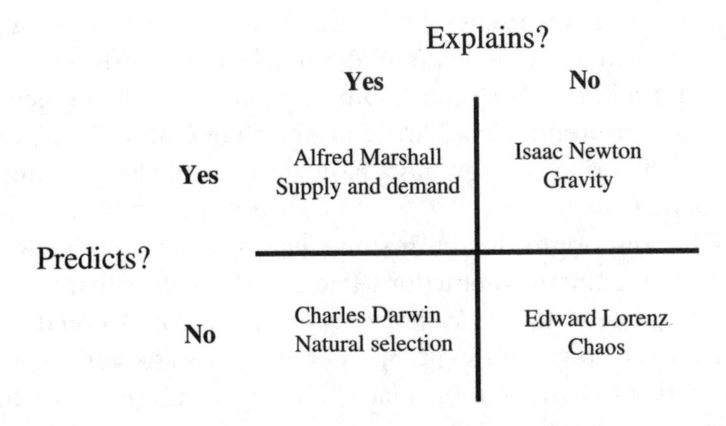

Figure 13.1 Theories Help Explain the Past and Predict the Future

Let's start with prediction. The upper-right quadrant displays Isaac Newton's theory of gravity, which predicts that the strength of attraction between two bodies depends on their masses and the square of the distance separating them. Newton could not explain how gravity works, but NASA still used this framework to land men on the moon and return them safely to earth. Prediction without explanation is at the heart of most analytics efforts. My old employer grew into a Fortune 500 firm by learning how to make predictions that came to pass even though the causal mechanisms remained unknown.

In contrast, the lower-left quadrant features Charles Darwin's theory of natural selection, which provides a causal explanation. The theory says that random genetic mutations better allow certain individuals to reproduce and pass along particular genetic traits to future generations. This theory explains why species evolve but does not offer a means of predicting future traits. Explanation without prediction is also informative. Studies of past pandemics do not help us predict future outbreaks but still offer helpful guidance.

The upper-left quadrant shows Alfred Marshall's price theory, embedded in classical economics. This framework shows how levels of supply and demand determine the price of a product at a point in time in a market. Marshall's theory may be used to explain the past (e.g., why an oil embargo brought a price increase) or predict the future (how increasingly available renewable energy sources could depress future oil prices).

A framework that offers both explanation and prediction becomes an invaluable addition to our analytical toolkit. Every successful businessperson has a basic grasp of the laws of supply and demand.

The lower-right quadrant shows chaos theory, a framework for thinking about interactions that are difficult to explain or predict. Meteorologist Edward Lorenz (1917–2008) noted how difficult it is to make long-range weather forecasts and explain why certain changes took place. His work showed how tiny changes in initial conditions bring unpredictable, unexplainable outcomes in future times and places.

Complexity associated with the so-called butterfly effect (where the mild atmospheric disturbance of a butterfly's wings brings a future tornado in a distant locale) humbles the smartest people on the planet. Frameworks in this quadrant represent theories under construction.

Even though no theory explains all phenomena in all situations, theories are useful as we find our way in the world. In my classes, I share a quote from philosopher Alfred North Whitehead (1861–1947), who remarked that the really useful training yields a comprehension of a few general principles with a thorough grounding in the way they apply to a variety of concrete details. Long after graduation, students will have forgotten the details but will remember how to apply principles to immediate circumstances. Psychologist Kurt Lewin (1890–1947) summarized this idea by saying that there is nothing as practical as good theory.

Figure 13.2 shows how theories are linked to material covered in this book. At the top is a parallelogram representing the theoretical plane of thought and at the bottom is another parallelogram representing the empirical plane of observation.

This relationship reminds me of Raphael's fresco *The School of Athens* displayed in the Vatican, which shows Plato pointing his right index finger to the heavens while Aristotle extends his right hand toward the viewer. Plato contemplates abstract ideas while Aristotle focuses on sensory experience. Figure 13.2 integrates the two approaches.

To take a simple example, let's explore the theory that large houses are more valued by home buyers. The theory connects the constructs of size and value. Constructs are ideas that are difficult to define precisely and are connected by proposition. Our proposition is that there is a positive relationship between house size and house value.

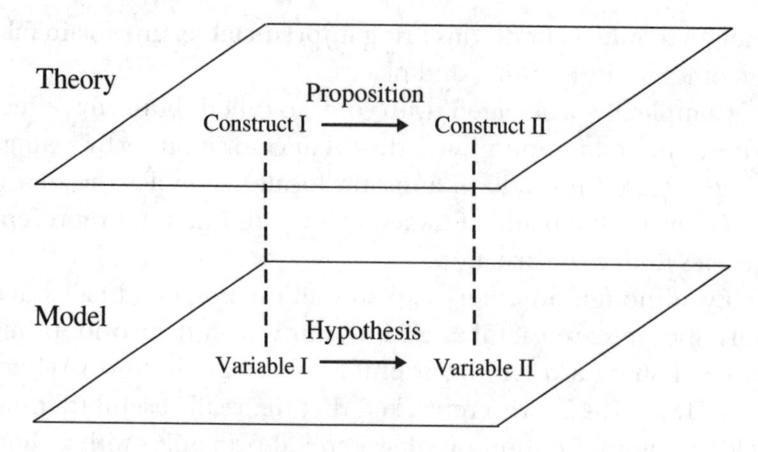

Figure 13.2 Models Bring Theories to the Real World

We create a model that ***operationalizes*** the theory by stating how constructs are measured as variables. Let's say that we decide to measure home size by the number of square feet of floor space. We select measurement rules that, among other things, exclude square footage associated with the garage, the unfinished utility space in the basement, and parts of the second floor where the ceiling height drops below five feet. Let's also say that we define value as the most recent sales price recorded in the land records of the local county Recorder's Office.

You're quick to identify all sorts of problems with how we defined our variables. Square footage does not capture ceiling height (e.g., many believe nine-foot ceilings offer more attractive living spaces than eight-foot ceilings), while the most recent sales price does not capture any side deals between sellers and buyers (e.g., a seller may have thrown in the washer and dryer to seal the deal, but this exchange was not reflected in the purchase price). As we discussed in Chapter 3, any effort to aggregate observations brings loss of information. Variables never capture the nuance of constructs.

A hypothesis (*H*) is a bridge between variables that creates a testable statement. We could say:

H: Square footage is positively related to home sales price.

There is no single, correct way to state a hypothesis. Different researchers craft hypotheses in varied ways. My experience is that studies go off the rails when people try to get too complicated. In this case, I've chosen wording that merely says that as the number of square feet of a house increases, so does its sales price.

We now get to the point of the chapter: How do we know if a hypothesis is true? Sorry to disappoint, but I don't believe we ever can. The best we may do is to rule out hypotheses by showing that we shouldn't believe them.

Popper says we could go on a swan hunt and note that all observed swans are white across hundreds of consecutive observations and yet fail to prove that our statement is true. The risk of a nonwhite swan lurking around the corner tempers any effort to say that we have proven anything. Finding one black swan, however, offers compelling evidence that we should abandon our old hypothesis and work to craft a replacement.

We show progress in science by ruling out hypotheses that are probably wrong. As we narrow the range of possible explanations, we focus increasing attention on those few guesses that continue to stand up to scrutiny. We test surviving hypotheses against new data sets to see if posited relationships still hold. In our home example, we could test whether a positive relationship between square footage and sale price continues to hold in different markets or time periods.

If we find a market where larger homes have lower sales prices, then we temper statements that our model is always and everywhere true. We should not necessarily jettison a model if we find that it does not always work. No physicist has yet created a theory of everything. We should be thankful when we find situations where a model does not work as expected, because the process of sorting through anomalous readings forces us to better understand our assumptions and data.

We should retain a flawed model if it remains useful. Over my career, I returned repeatedly to the efficient markets hypothesis, capital asset pricing model, and Black–Scholes option pricing model – frameworks that I knew had problems but nevertheless offered useful conclusions.

In this journey of hypothesis generation and testing, we seek to avoid ***confirmation bias***, the tendency to interpret data in a way that corroborates preexisting beliefs. The classic example is the sharpshooter fallacy, where we fire bullets at a wall and then draw a bull's-eye around the bullet holes: we created and confirmed an argument using the same set of data.

Finding ways to cast doubt on what we believe to be true is healthy. Inadequate skepticism may cause us to hold on to hypotheses that are wrong and make horrendously bad decisions based on them.

A solution is to modify a proposed relationship between variables so that it is expressed in the form of a ***null hypothesis***, using the signifier H_0. A null hypothesis is a testable statement that says nothing interesting is happening. If we gather evidence to disprove H_0, then we have a basis for believing that something is indeed afoot.

Suppose you are the mayor of a safe, small town that experiences little crime. A local shopkeeper calls the police to report that someone broke into his store and stole some inventory. Later, a patrolman apprehends a stranger who is not from the area. A believable hypothesis is:

H: The stranger stole the goods.

The local merchants are up in arms and want the police to lock up the stranger. Yes, you note, there was no reported crime before the outsider arrived, so there is some basis for concluding that he did it. Yet, you vaguely remember from school the principle that Y following X does not mean that X caused Y, a logical fallacy expressed in Latin as *post hoc ergo propter hoc*. You feel queasy when considering the merchants' request.

You consider other possible explanations for what happened. Perhaps some high school kids broke in as part of a prank, the financially strapped shopkeeper arranged the theft to file a bogus insurance claim, or his ex-wife took the property as retribution for a fractious divorce settlement.

If the merchant has it wrong, then convicting the outsider will bring a whole new set of problems. We want to bring resolution without an inappropriate conviction. To do so, we use the null hypothesis to mitigate the risk of confirmation bias. We say:

H_0: The stranger did not steal the goods.

We then have a trial in which the city prosecutor attempts to convince a jury to reject the null hypothesis while the defendant's attorney attempts to highlight the weaknesses of the prosecutor's case. In the end, the jury decides whether the prosecutor has marshaled sufficient evidence to disprove the null hypothesis that the guy didn't do it.

Figure 13.3 shows the reasoning of invoking the null hypothesis. The top panel puts forward the testable statement in its direct form. Should we find evidence to support the hypothesis, then we could decide to accept it. The problem with this approach is that we may ignore inconvenient facts that belie the hypothesis. Perhaps a witness saw someone resembling the defendant in a different part of town when the break-in occurred.

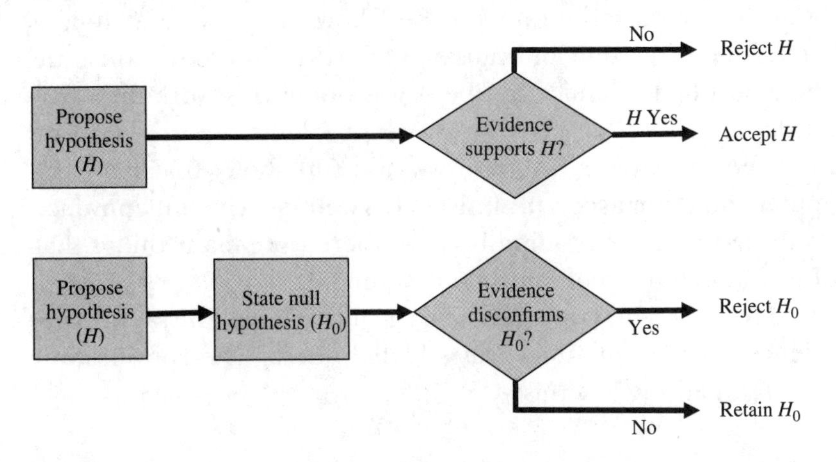

Figure 13.3 Use of Null Hypothesis Mitigates Confirmation Bias

The bottom panel adds a speed bump. Introducing an additional step forces us to ask if we may reject what we don't believe to be the case. We consider the evidence and then determine whether it is sufficient to persuade us to reject H_0. If so, we rule out the idea that nothing is going on and conclude that there is some type of relationship between the variables in our hypothesis. Note how this conclusion is more modest than saying that we accept the original hypothesis.

If we fail to gather sufficient evidence to falsify H_0, then we hold on to it (and concede that we can't say that something special is happening) as we go about our business. Retaining H_0 prevents us from overreaching and taking too much credit in our ability to explain the past or predict the future.

Scientists believe the slower, more cumbersome path in the bottom panel is the better approach to finding our way in the world. Using the null hypothesis mitigates the risk of confirmation bias and helps prevent us from doing harmful things based on an analysis of inconclusive statistical tests.

An example of ruling out hypotheses comes from my experience of having trouble logging on to my university's website. Figure 13.4 shows a screenshot from my computer on the evening of October 7, 2017. Just to add a note of dramatic irony, note that the far-left tab on my browser shows that I was watching a video on correlation and causation. Further, just to demonstrate how boring academics can be, I was doing this work on a Saturday night.

The display for the tab www.case.edu shows that my computer could not access my university's website. The unhappy face reflected my state of mind because there were many things that I had hoped to accomplish that evening.

I rebooted my computer, tried again, and had the same experience. I called the university's IT help desk. The conversation went something like this:

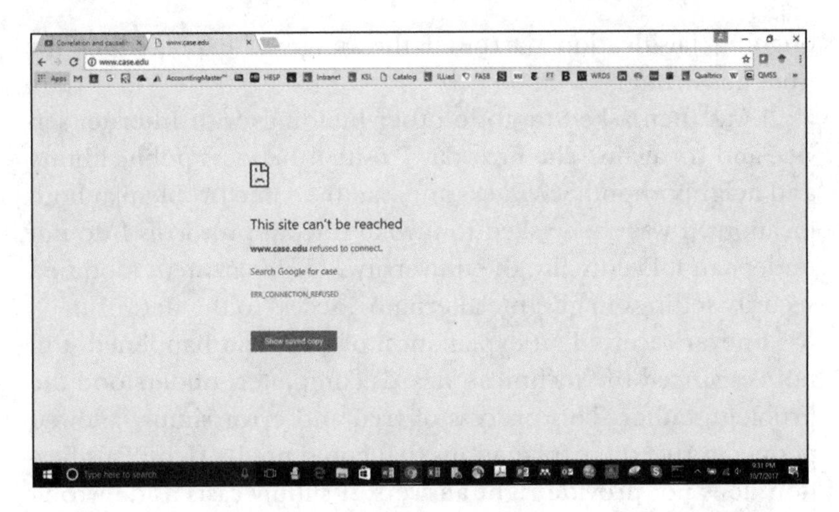

Figure 13.4 Screenshot of My Computer on 7 October 2017
Source: Google LLC

Professor King, is your computer turned on?

Yes, I can see the illuminated screen right now.

Did you turn off the computer and restart it?

Yes. Tried that but I still have the same problem.

Is your Internet connection down?

No. The Wi-Fi icon shows that I have a strong connection.

Can you access other websites?

Yes. I just visited the websites for YouTube and the *New York Times*.

Can you access our website on a different browser?

No. I'm on Chrome and also tried Safari and Explorer. Same result.

You get the idea. The representative was throwing out null hypotheses, and I was ruling them out. This process of ongoing falsification narrowed the range of explanations for what was wrong.

I was then asked to go to other buildings with Internet service and try again. The next day I visited the local public library and neighborhood Starbucks and had the same problem at both locations. I was then asked to invoke tracing protocols I do not understand. Eventually, the university's IT department modified security settings to give me alternative access to the network.

I never received an explanation of what had happened. I'm not convinced the technical wizards completely understood the problem, either. This process of trial and error simply allowed people to rule out explanations that could not be right. Falsification does not provide right answers; it simply casts aside wrong ones until we're left with a surviving explanation that seems to do the job. Falsification is an example of pragmatism, a search for something that works without attempting to find ultimate truth, the topic of Chapter 17.

Back to our break-in story. Suppose that the town prosecutes the stranger for the break-in. The jury weighs evidence provided by the prosecuting and defense attorneys and is not persuaded to reject the null hypothesis that the guy didn't do it. The police free the man. The trial did not answer the question of who committed the crime; it simply ruled out one of many possible explanations.

The cumbersome process of setting up a hypothesis, expressing it in null form, and then seeking to falsify the null slowed things down so that there was no rush to judgment on flimsy evidence. Falsifying null hypotheses serves society by reducing the likelihood that we bring harmful change based on flawed data or reasoning.

Falsification is not a panacea. It is always possible to get things wrong. Juries have wrongfully convicted innocent defendants and have failed to convict those who have committed crimes.

Figure 13.5 provides a framework for thinking about mistakes from statistical reasoning. We begin at the left and look

		Unobservable Reality	
		Nothing happening (H_0 is True)	Something happening (H_0 is False)
Statistical Conclusions	Nothing detected (Retain H_0)	Okay	False Negative Type II error [β]
	Something detected (Reject H_0)	False Positive Type I error [α]	Okay

Figure 13.5 Falsification Brings Two Kinds of Problems

across the two rows. We are flawed human beings who walk the earth with limited understanding of the universe around us. We have statistical tools to sort through William James's metaphor of blooming, buzzing confusion brought by events unfolding around us.

Armed with the tools of falsification and the null hypothesis, we use evidence and reasoning to assess whether we've detected nothing interesting going on and retain the null hypothesis (i.e., we conclude the stranger was simply passing through town without stealing goods) or we've detected something unusual, reject the null, and conclude that something significant happened (i.e., he broke into the store and stole the stuff).

I cannot emphasize enough that our decision to retain or reject H_0 does not turn on a number delivered by a quantitative statistical measure such as *p*-value, covered in the next chapter. Our conclusion rests on whether our evidence is sufficient to persuade us to reject the null and conclude that something significant has been detected from our analysis. If our judgment leads us to believe that we failed to find anything interesting, then we hold on to the null hypothesis and refrain from taking action.

Alas, to paraphrase Oliver Cromwell, we could be wrong. We will almost never know the truth behind statistical inference because humans rarely have access to all the data. If we go to the top of Figure 13.5 and look down the columns, we share the perspective of the gods (note the lower case *g*) on Mount Olympus staring down at miserable humans who walk the earth's surface. The gods have a broader perspective, see things we cannot, and know whether the null is true or false. They know whether the stranger committed the crime but won't tell us the answer.

We now have four possibilities shown by this matrix. In the upper left, our statistical analysis got us to the right answer. We did not find anything interesting and retained the null (or, as some statisticians say, we failed to reject H_0) when the guy didn't do it. Justice was served when we let him go. In the lower right quadrant, we were persuaded to reject H_0 and convicted the stranger when H_0 was really false. Justice was again served when we locked him up.

We run into problems on the other diagonal when we get it wrong. In the lower left, we have the serious problem of deciding that something is going on and rejecting the null hypothesis when nothing is afoot. This is a ***false positive***, also known as a Type I error or α. The tests we used suggested that something unusual is happening when nothing special is going on. Here we convict an innocent man, which creates all sorts of downstream problems for society.

In the upper right quadrant, we were not persuaded by the evidence to reject H_0 and chose instead to retain the null. We let the stranger go when he did indeed do it. This is a ***false negative***, also known as a Type II error or β. In a false negative the statistical tests failed to identify something interesting going on.

Generally, researchers are less worried about false negatives than false positives because we would rather fail to intervene than to intervene and do more harm than good. A way to remember this to reflect on the physicians' Hippocratic Oath requiring doctors to, first, do no harm.

Because we can never have the perspective of gods on Mount Olympus, we must always live with the idea that statistical reasoning brings the risk of mistakes. We can never completely rule out the possibility of Type I and Type II errors. In fact, the harder we push to drive down the incidence of false positives, the greater our risk of experiencing false negatives.

In management research settings, a commonly accepted level for false positives is 5% of trials ($\alpha = 5\%$) and a commonly accepted level for false negatives is 20% of trials ($\beta = 20\%$). In Appendix B, we show how combining these facts could lead to a situation where a third of all scientific discoveries are wrong.

Recap of Chapter 13 (Null Hypothesis [H_0])

- Hypotheses propose relationships between variables.
- Hypotheses expressed in null form (H_0) state that not much is going on between variables.
- Science uses evidence and reasoning to attempt to disprove null hypotheses.
- Using the null hypothesis mitigates risk of confirmation bias and of intellectual overreach.
- Statistical tests are unable to prove that a hypothesis is true.

14

p-VALUE [p]

In the previous chapter, we showed how scientists seek to mitigate confirmation bias by expressing a hypothesis in null form and then searching for evidence to disprove a bland statement that nothing interesting is afoot. A common question to ask is what sort of evidence is helpful to falsify null hypotheses. This chapter introduces ***p-value***, a statistical measure used for falsification.

A p-value offers a probability that there is no significant difference between a measured variable and a benchmark value associated with little interesting going on. A small probability suggests that something interesting is happening. The most important thing to take away from this chapter is that low p-values offer support but not proof that we may reject null hypotheses.

Suppose you walk into my classroom and I show you two pennies. The first is a shiny coin minted last year. The second is 30 years old, produced before any students in the classroom were born. Inspection shows that the older coin, dulled with age, has nicks and scratches that have accumulated over a generation.

Table 14.1 Results of Flipping Two Coins for 100 Trials

	Old coin		*New coin*	
	#	%	#	%
Heads	52	52%	47	47%
Tails	48	48%	53	53%
Total trials	100	100%	100	100%

I ask if the two coins are identical. Of course not, the students respond. I then ask if the two coins are significantly different. Small conversations emerge. *Well, Professor King, what do you mean by significantly different?*

Great question. I define significantly different as having unequal likelihoods of coming up heads after coin flips. If the two coins are "fair," then about half of all flips will result in heads for each coin. Yes, I did look this up, and there is an academic paper that says it's safe to assume that there's a 50% chance of flipped coins landing as heads (Diaconis et al., 2007).

So, what I'm really asking is whether the accumulated nicks and scratches have changed the older coin's physical properties so that it's no longer fair. This question moves us from the realm of probability, with outcomes governed by known probability distributions, into the world of statistics, with unknown outcomes that are difficult to measure. My question is about uncertainty, not risk. I now have the students' attention.

I then share previously computed data, shown in Table 14.1, because we don't have time to watch and tally 200 coin flips.

Were the results identical? Absolutely not. The old coin had 52 heads while the new coin had 47. May we conclude that the old coin is significantly different from the new one? The students confer, murmur, and say probably not. They say that the answer is not clear because we are dealing with the luck of the draw. As we discussed in Figure 3.1, there is always a risk that inferences based on samples lead to flawed conclusions.

So, here is one way that a numerate person could assess whether anything is afoot. First, we offer a hypothesis that something interesting is going on:

H The old coin and new coins produce significantly different numbers of heads.

We then use the null form to mitigate the risk of confirmation bias so that, if we make a mistake in our statistical reasoning, we're less likely to do something stupid. One way to express the null hypothesis is to say:

H_0 The old and new coins do not produce significantly different numbers of heads.

We now ask if we have sufficient statistical evidence to reject the null. One statistical tool in a propeller head's arsenal is the comparison of proportions test. This tool assesses whether proportions between two samples are significantly different.[1] The calculator uses proportions of two samples (here, 52% heads for the old coin and 47% heads for the new one) plus the number of observations in each sample size ($n = 100$ for each sample) to deliver statistical output that includes a *p*-value.

In this instance, $p = 0.4806$. I interpret this figure as saying there is a 48% chance that there is no significant difference between the likelihood of the two coins delivering heads. We learned in Table 2.5 that numbers are meaningless in isolation. It is only after we enforce comparisons that information may be extracted from data.

I choose 5% as my standard for statistical significance. This is the same idea as $\alpha = 5\%$, introduced in the previous two chapters. A 5% level of uncertainty means that I'll allow myself to be

[1] I used https://www.medcalc.org/calc/comparison_of_proportions.php to obtain *p*-values for this example (retrieved 26 January 2021).

persuaded that something is going on if there is only a 1-in-20 chance of getting things wrong by reaching a false positive.

If the stakes are high – and the consequences of having a false positive are really bad – then I may choose to reduce α to a lower figure of, say, 1%, which is associated with a 99% confidence level. If making a Type 1 error would be catastrophic, then I would demand a super-low α (perhaps 0.1%) as the benchmark to reject the null hypothesis.

Since we're just dealing with coin flips (and wagers that confer little more than bragging rights), I readily accept 5% as my benchmark. Remember, rejecting the null means we think something out of the ordinary is happening and accept that some type of intervention is warranted.

I now compare $p = 48\%$ with my reference point of $\alpha = 5\%$ to note that the *p*-value is much larger. At $p = 48\%$, there is a 48% chance that there is no significant difference between the fraction of heads delivered by the old and the new coins. I find 48% to be a high value and a long way from my 5% benchmark. I do not have sufficient evidence to be persuaded to reject the null hypothesis. I choose to retain H_0 and believe that the old and new coins do not produce significantly different numbers of heads.

Although this did not happen, suppose that I flipped both coins 100 times and observed 64 heads from the old coin and 47 heads from the new one. The statistical software now shows that $p = 1.6\%$, suggesting there is just a 1.6% chance that there is no significant difference between the two coins.

Here, the *p*-value is lower than my 5% benchmark for α, and I choose to reject H_0. I conclude that the old and new coins deliver significantly different numbers of heads. Of course, I know that I'm still subject to Type I error where nothing is happening, even in the face of something identified by my statistical test. However, I'm willing to live with this contingency because the most I will lose is some low-grade bets to a group of impoverished students.

Let's pause to emphasize two points. First, when *p*-values are larger than our selected value for α, then all we are saying is that results appear to be due simply to chance; nothing interesting seems to be going on.

Second, a *p*-value below α does not mean that the statistics police force us to reject the null hypothesis. We may rightfully remain unpersuaded in the face of a blisteringly low *p*-value because, say, we worry that our model omits other important variables, our sample is biased, our variables do not measure constructs contemplated in our theory, or our measurement processes are not reliable. We use statistics to reduce uncertainty, not to prove things.

Permit me another example to provide a visual explanation of how α and *p*-value are related. I offer stylized data based on a real event.

Many college students are not morning people and choose to enroll in courses that start after 10 a.m. The COVID-19 pandemic forced some international students to take courses from their home countries. To accommodate students living in the Middle East and Asia, we moved certain classes to the early morning to provide synchronous, remote instruction for all students.

Among my concerns was whether an earlier start time would interfere with student learning. In one course that I teach, the mean grade for a particular assignment had hovered around 88% with little variance. I formed the following hypothesis:

H Student grades will be lower when the class start time is early in the morning.

As the semester unfolded, I noted that the mean grade for that assignment was 85%. Considering myself numerate, I knew enough to realize that I am at risk of suffering from confirmation bias, where I interpret data in a way that corroborates preexisting beliefs. To show that I eat my own cooking, I expressed my hypothesis in null form:

H_0 Student grades will be unaffected by class start time.

I then set about gathering evidence in efforts to disprove H_0.

A nagging worry was that the variable of mean grade may not be a valid measure of the underlying construct of student learning. Learning and grades do not always go hand in hand.

Consequently, I decided ahead of time to avoid making any sweeping generalizations from this exercise, regardless of what is discovered. This line of reasoning is addressed in Chapter 17.

The sample statistics were $n = 31$, $x = 85\%$, $s = 5.5\%$. Leveraging what we've covered earlier in the book, we note that the sample size crosses the informal threshold of 30 observations, so we may take some assurance that the sample size is sufficiently large to make a defensible inference. The coefficient of variation (s/x) is below 7%, so, as discussed in Table 5.2, we have a green light to consider using x to make predictions or explanations. It looks like something is different, but let's run the numbers harder.

We then ask if the current semester's class is representative of the population of all students who take the course. Any type of biased sample could, to use a British metaphor, throw a spanner in the works. My cursory review of the student roster and reflection on my interaction with students leads me to believe that this year's class is not appreciably different from those in prior years.

Starting with a desired confidence level 95% ($\alpha = 5\%$), using the central limit theorem (and assuming that our data may be considered a random sample of 31 trials), and consulting Table 12.1, we see that a margin of error for an inference should be about plus or minus two standard errors. Thus, our inference for the mean grade for the population of students taking the course at the earlier start time is roughly:

$$\begin{aligned} \mu &\approx x \pm 2(s/\sqrt{n}) \quad \text{@ 95\% confidence level} \\ &\approx 85 \pm 2(5.5/\sqrt{31}) \\ &\approx 85 \pm 2(1) \\ &\approx [83, 87] \end{aligned}$$

This confidence interval says that we have some evidence to suggest that the unobservable population mean grade of all students taking the course at the earlier start time would fall within the range of 83 through 87.

I notice that the upper end of the confidence interval falls below 88, the ***critical value*** (CV) that serves as our benchmark for a test of statistical significance. Put another way, the confidence interval does not capture the CV.

We have some statistical evidence for saying that something is going on. If the CV had fallen within the confidence interval, then we could have said that the outcome was likely due to chance, because 95% of all test scores should be expected to fall within this range. We would have little evidence to reject the null hypothesis that nothing special is going on.

Our test, however, showed that the CV of 88 fell outside the two-standard-error confidence interval spanning the sample mean. I muster up the courage to reject the null hypothesis and suppose that something is afoot. If things were just left to chance, there would be a less than 1-in-20 probability of us seeing the assignment grade being this low.

Figure 14.1 presents the situation graphically. The *x*-axis shows the variable we care about – the average grade for the

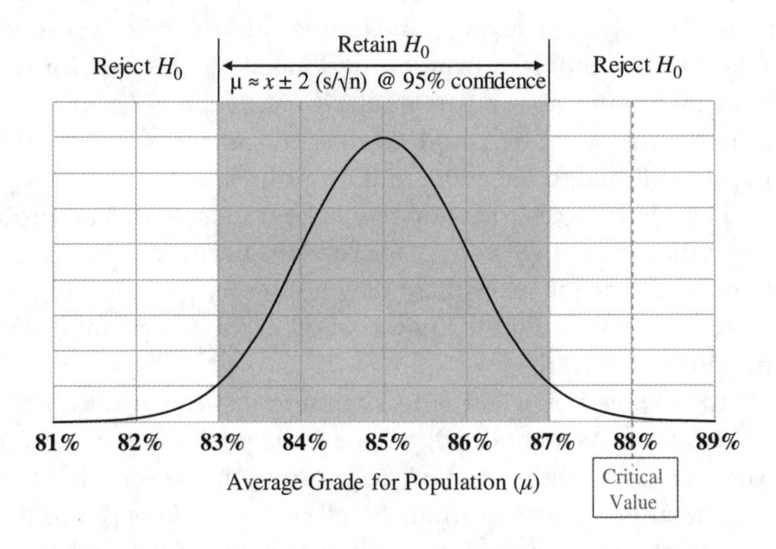

Figure 14.1 Determining Whether Critical Value Falls Within Confidence Interval

assignment given to our class at the new time. The midpoint, 85%, is the sample mean for that class. The *y*-axis shows relative frequencies for different levels of grades.

The bell-shaped curve is an estimated histogram showing how likely the true, unobservable population mean (μ) is to occur at different grade levels. Here, μ is the signifier for the idea of the average grade of all students who would take our course at the earlier start time. We remember from Chapter 6 that a bell-shaped curve has an identical mean, median, and mode, that the area under the curve adds up to 100%, and that the curve is symmetrical with predictable tails. Our best guess is that $\mu = 85$, as shown by the peak of the curve as this value.

If we decide to use the central limit theorem (CLT), assume that our sample is unbiased, and set α equal to 5%, then the shaded area that is plus or minus two standard errors represent the confidence interval likely to capture μ with a 95% confidence level. The CV is 88, outside of the shaded area associated with the confidence interval.

Because the CV is outside the confidence interval, we have a basis for rejecting H_0 and concluding that the 88% test score is significantly different from our estimated mean grade for students taking the class at the earlier time. If the CV had fallen in the shaded area, we could say that this outcome was not sufficiently unusual to reject the null hypothesis.

This chart looks great and is easy to understand. The problem is that it takes up a lot of space. Statisticians choose to not devote the space to show these diagrams every time they test an inference. All they want to know is whether the critical value falls within the shaded area.

An accepted shortcut is to simply report the *p*-value for a statistical test. As mentioned, *p*-value is the probability that there is no significant difference between two numbers – in this case the 85% inferred average grade for all students taking the course at the early start time and the 88% average grade recorded for students who had taken the course at the customary start time in previous semesters.

If we had defined our exercise as a "one-tailed" test and sought to calculate the area under question to the right of the CV, then our software would return a value of $p = 0.13\%$. This means that the area of the curve to the right of 88% is the ratio of 0.0013 to 1, or about 1 in 769. This is a small percentage.

An interpretation is that the likelihood of μ falling to the right of 88% is incredibly unlikely given a sample with statistics of $n = 31$, $x = 85\%$, and $s = 5.5\%$. A low *p*-value suggests that μ (the unobservable population mean for all students taking the course at the earlier start time) is significantly different from 88% (the average grade for all prior students who took the course at the later start time).

Our estimate for μ seems to be meaningfully separated from the critical value. We don't know the precise value for μ, but it sure looks like it is lower than 88%. Some people would feel that there is sufficient evidence to reject the null hypothesis that student grades are unaffected by an earlier start time.

I caution that going from statistics to a story is a big intellectual leap. There may be other reasons for the lower grade besides changing the class start time. Among many other explanations is the possibility that students were suffering from the mental strain associated with remote learning, which brought down the average grade. Or, I was a less effective teacher during the semester in question because I was teaching over the computer. In Chapter 16, we discuss how ruling out alternative explanations is a necessary condition for making a causal argument.

Using the same data, we could define our exercise as a "two-tailed" test and calculate the area under the curve above 88% and below 82%. The area under these two, symmetrical tails conveys the likelihood that the μ is significantly removed from the critical value without concern to direction.

Because a normal distribution is symmetrical, the combined area for a two-tailed test is twice what it is for a one-tailed test ($p = .26\%$ for the two-tailed test; $p = 0.13\%$ for the one-tailed test). The two-tailed test answers the question of whether the average grade for the population of students taking the class at the new start time is significantly different from that of students who took the course at the later start time. Users of two-tailed tests are indifferent as to whether one number is higher or lower than the other; rather, the user simply worries about the two being unequal.

My experience is that practitioners care only if one group is different from another group. A two-tailed test delivers a twice-as-large p-value and is thus a more conservative test of differences. When in doubt, use a two-tailed test.

So, to summarize, p-value is a handy way to ask whether one number is significantly different from another. Our definition of significantly different turns on our selection for α. If we set $\alpha = 5\%$ and we find that $p = 4\%$, then we have some justification to say that the two numbers are meaningfully different. We could choose to reject our null hypothesis and conclude that something is going on. If we find that $p = 6\%$, then we could say that our stats have not met our before-the-fact standard for statistical significance. In this case, we could retain our null hypothesis and say that we have been unsuccessful in efforts to falsify our null hypothesis.

Of course, you are being paid to use your judgment, not to mindlessly follow static decision rules. Please do not be seduced by the elegant simplicity of comparing p with α. Too many people view such an exercise as providing "the answer" to a statistical test. I could not disagree more.

If you don't believe me, consider reading *The Cult of Statistical Significance: How the Standard Error Cost Us Jobs, Justice, and Lives* (Ziliak & McCloskey, 2008). This 300-page manifesto makes clear that p-value is nothing more than an imperfect tool to sort through the messy world around us.

As discussed in the following chapters, selecting an α of 5% was arbitrary, we made all sorts of simplifying assumptions when invoking the CLT, and our answer ($p = 0.26\%$) does not prove anything. The differing grades may have resulted from something other than a change in starting time.

One important caveat when reviewing p-values is to consider the number of observations that went into the calculation of standard error. If we had a big data set, say n = 10,000 observations (something common in accounting databases for large organizations), then the standard error (calculated by dividing s by the square root of n) will be quite small.

Small standard errors bring narrow confidence intervals. Put another way, the risk of false positives increases when working with huge datasets because it gets easier for a critical value to fall outside confidence intervals developed from skinny standard errors.

A final topic to consider is my response to a common question of how much evidence we need to be persuaded. My response is that it depends on prior beliefs. If a client has a deeply rooted point of view, it will either take an overwhelming amount of evidence to bring about change or be downright impossible to do so. Put simply, don't get mad if your analysis fails to influence others.

Suppose that I tell you that I have a superpower: the ability to predict coin flips. You're skeptical but are willing to invest time to test the null hypothesis that I have no such ability. We flip a coin five times in a row, and I correctly call heads for each trial. The probability of me doing this is 1 in 32. If we compare the 3.125% chance of getting this right with a customary level of an alpha of 5%, we could justifiably reject the null that I do not have any special predictive powers.

I am confident that you would choose to retain the null hypothesis. Your prior belief is that no one can predict fair coin flips. You conclude that either I'm lucky or have a loaded coin. I could call 25 straight coin flips in a row (an almost impossible task), but it's likely that you would remain unpersuaded.

A real-world example is reaction to Special Counsel Robert Mueller's investigation of whether the Russians colluded with the Trump campaign in the 2016 presidential election. Mueller's team had a federal mandate to spend as much time and money as needed to sort through the issues at hand. In your career, you will never have the time and resources afforded the Mueller probe.

After publication of the final report, little was concluded. Trump critics read the report and found evidence of election tampering while Trump advocates read the same document and felt that he was exonerated. My assessment is that these two groups held such strong prior beliefs that no amount of evidence would have been sufficient to persuade members of either side to change their minds.

Recap of Chapter 14 (*p*-value [*p*])

- *p*-value provides a probability that one number is statistically different from another.
- If a *p*-value is low (say, below 5%), then we have evidence that two numbers are different.
- Many statisticians use low *p*-values as evidence to reject null hypotheses.
- Low *p*-values are easy to use but don't prove anything.
- Prior beliefs limit the ability of statistical evidence to change people's minds.

15

SLOPE $[r(s_Y/s_X)]$

In the previous two chapters, we noted how theories may be developed to predict or explain phenomena in the world around us and how models may be created to test those theories. The tools of falsification and *p*-value, now part of our numeracy toolkit, help us evaluate models.

This chapter builds on these ideas to suggest how we may use models to make informed predictions. The subsequent chapter suggests how we may use models to make explanations. The summary point distinguishing the two chapters is that prediction is about using correlation, whereas explanation is a more ambitious goal requiring the use of a three-part test.

Chapter 8 explained how correlation, *r*, is a means of measuring the strength and direction of a linear relationship between two variables. A value approaching +1 suggests that increases or decreases in variable 1 will, respectively, be associated with increases or decreases in variable 2. A value of *r* approaching –1 suggests that increases in variable 1 are associated with offsetting decreases in variable 2, and vice versa.

Chapter 9 introduced the idea of falsification through the use of r^2, also known as the coefficient of determination. By squaring the correlation coefficient, we may estimate how much of the movement of one variable may be explained by movement in another. Calculating the complement, $1 - r^2$, shows how much movement in one variable cannot be explained by movement in another. A value of $1 - r^2$ approaching 100% suggests that one variable is a lousy predictor of another, at least if we're searching for a linear relationship between the two.

While r and r^2 are helpful, they do not tell us how much movement in one variable is associated with movement in another. This is where we introduce the concept of ***slope***, a measure of how much we could expect a second variable to move from a one-unit change in the first.

The notation $r(s_Y/s_X)$ signifies the slope of a linear relationship. This expression says that slope is equal to the product of the correlation coefficient times the ratio of the standard deviation of the second variable (s_Y) divided by the standard deviation of the first variable (s_X).

The intuition of this is that we need a way to calculate the ratio of change in Y from a change in X. The correlation coefficient, r, shows the degree that two things move together, but this statistic has no unit of measure. Standard deviations, s, show variance and are expressed in the same units as the observations in a sample.

Dividing s_Y by s_X gives a unit of measure for a slope (change in Y given a change in X, as in change in oranges given changes in apples). Multiplying this ratio by r scales the correlation coefficient so that it is no longer a unitless measure bound by the interval –1 to +1. We did this exercise to show slope of 0.68 oranges per apple in Panel A of Figure 9.1.

Table 15.1 applies this idea to the data set shown in Figure 9.2. The sample shows six observations of house size, measured in square feet, and sale price, measured in U.S. dollars.

Table 15.1 Sample Data and Statistics for Home Sales

	X	*Y*
	House Size (square feet)	Sales price (U.S. dollars)
	1,800	239,000
	3,800	364,000
	3,000	328,000
	2,750	315,000
	4,100	512,000
	3,450	500,000
Standard deviation [s]	827	108,452
Correlation coefficient [r]	0.837	
Coefficient of determination [r^2]	70.0%	
Unexplained variation [$1 - r^2$]	30.0%	
Slope [$r(s_Y/s_X)$]	109.7	

The sample statistics at the bottom include the standard deviation of each variable, the correlation coefficient between the two, and the slope of how much the *Y* variable moves in response to movement in the *X* variable, assuming a linear relationship between the two.

To use terminology introduced in Chapter 13, our two constructs are value and size. Our theory is that home value is positively related to the size of house. We operationalize these constructs using the variables of sales price in U.S. dollars and home size measured in finished square feet. The sample data come from recent transactions captured by our local county Recorder's Office.

If you start getting huffy about my theory, variables, and sample frame, then I take comfort that you've been paying attention. We could drive a truck through the holes in my theory, how we

Table 15.2 Simplified Regression Output for Home Size and Price Data

Regression Statistics	
Multiple *R*	0.837
R Square	70.0%
Adjusted *R* Square	62.5%
Standard Error	66,442
Observations	6

	Coefficients	*Standard Error*	*t Stat*	*P-value*
Intercept	30,800	116,377	0.26	80.4%
Square feet	109.7	35.9	3.05	3.8%

operationalize it, and how we collect data to test it. A little voice in your head should be screaming at you about all sorts of issues ranging from measurement ***validity*** and ***reliability*** to sample size and bias. Just stick with me for purposes of keeping this simple.

Plugging in the numbers, we see that $r(s_Y/s_X) = 0.837$ (108,452/827) = 109.7 dollars per square foot. This measure says that a one-square-foot increase in home size may be expected to bring a $110 increase in sales price, assuming that there is a linear relationship between these two variables.

An easier way to get at this relationship is to use the statistical tool of linear regression. To keep this chapter short, I'll skip the description of what regression analysis does and how it does it. There are many online videos covering this topic.

I performed a regression analysis in Excel, simplified the output, and made some formatting changes to keep this discussion short. Results are shown in Table 15.2.

The power of regression analysis is shown at the bottom of the table. The dependent or *Y* variable (sales price in dollars) has been regressed against the independent or *X* variable (square feet).

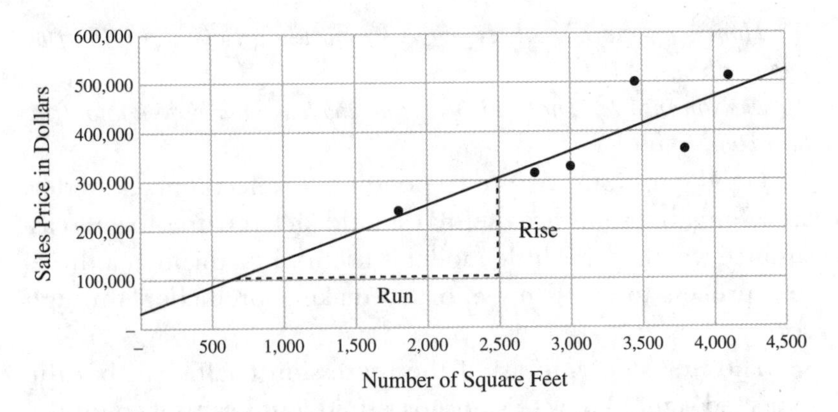

Figure 15.1 Graphical Representation of Linear Regression Between Sales Price and Size

The slope of the line connecting the ordered pairs of dots is 109.7 dollars per square foot, shown under the Coefficients column. The *y*-intercept for the relationship is 30,800 square feet. Figure 15.1 presents this relationship in graphical form.

The dots plot the six observations. The solid black line graphs the linear relationship of $y = 30{,}800 + 109.7$ (square feet). Note how the line crosses the vertical axis on the left at \$30,800 and then extends to the upper right at a rate of \$109.7 per square foot. This slope is the ratio of rise (the increase in value of the dependent variable) over run (the increase in value of the independent variable).

What makes regression useful is the ability to make predictions about things that have not been observed. For example, we may contemplate selling our home and moving to a community with less snow and warmer temperatures. Our home size, as shown in an online data source, is 3,749 square feet, a value not represented in the sample data. If we make the assumption that our house is like those appearing in our sample, then we may plug our home size into the equation to obtain a prediction:

$$\text{Predicted sales price} = \$30{,}800 + (\$109.7)(3{,}749) = \$442{,}065$$

Honey, my quick analysis suggests we could get about $440k if we sold our home.

Are you insane? That number is way too low. How do you know this prediction is any good?

Good questions. Before responding, I reflect that all models are wrong, but some are useful. I should kick the tires to get reassurance whether this little model is useful. I go through a three-step process to assess my effort to make a prediction through correlation.

The first step is to ask if the regression coefficient has the expected sign. I know this sounds stupid, but I cannot count the number of times that I have been surprised by direction of correlation. Being numerate means keeping an open mind when looking at statistical output. Here the regression coefficient of 109.7 is positive, suggesting that increases in home sizes are associated with higher sales prices. This result is consistent with my expectation, so I continue.

If I had found that larger home sizes were associated with falling sales prices, then I would have to ask questions about the reasonableness of my theory, the validity and reliability of my variables, the nature of my sample, and so on. The final chapter of this book offers a process for such tire-kicking.

The second step is to ask whether the regression coefficient is just a fluke. This is where we use the tools of falsification and *p*-value. Let's take the null hypothesis for our model:

H_0 There is no linear relationship between square feet and sales price.

How could we show that there is an absence of a linear relationship between these two variables? Well, we could test whether the slope of the line is statistically different from zero. If it is, we have evidence to rule out the null hypothesis.

A slope of zero suggests that movement in one variable has no influence on movement of another. For example, consider this stylized example of the sales activity of a lemonade stand

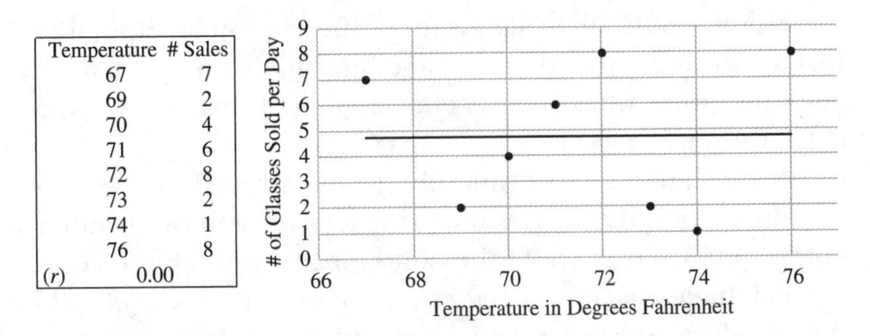

Temperature	# Sales
67	7
69	2
70	4
71	6
72	8
73	2
74	1
76	8
(r) 0.00	

Figure 15.2 Correlation Coefficient of Zero Suggests Absence of Linear Relationship

over eight days, presented in Figure 15.2. The data in the left panel show average daily temperature in degrees Fahrenheit and the number of glasses of lemonade sold.

One way to assess whether a relationship is present is to run a quick correlation coefficient calculation. We see that $r = 0$ and that the best-fit line connecting the dots is flat (or has a slope of zero). Increases (decreases) in temperature do not appear to be associated with increases (decreases) in lemonade sales. We probably should not rely on temperature to predict sales in our lemonade stand.

Should I show this analysis to anyone, I would highlight that I chose to use Pearson's r to calculate the correlation coefficient between temperature, which measures continuous, interval data, and sales, which counts discrete, ratio data. Earlier in the book we noted that r makes use of subtraction and division and that there are problems using division on interval scales and subtractions on discrete data. Highlighting limitations in our analysis builds trust with skeptical audiences and calms propeller heads who may worry that we're seeking to win statistics prizes.

The key point is that a correlation coefficient near zero offers evidence that there is an absence of a linear relationship between two variables. We may use this fact to set the critical value at zero for a p-value calculation. We could then provide evidence to falsify our null hypothesis if we may show that the regression coefficient is significantly different from 0. If so, then we have evidence that there is something going on between size and sales price, which bolsters our argument that the two variables are correlated.

Regression output provides the *p*-value needed to assess the probability that the regression coefficient is significantly different from 0. Going back to the bottom panel for Table 15.2, we see that the *p*-value for the regression coefficient (slope) of 109.7 dollars per square foot is 3.8%. This says that, if we choose to use the CLT, then there is only a 3.8% chance that 109.7 is not statistically different from the critical value of 0.

If we set α = 5%, then our *p*-value is below our benchmark, and we have evidence to reject the null hypothesis that there is no linear relationship between sales price and square feet. Again, I emphasize that low *p*-values don't prove anything.

Figure 15.3 shows this relationship graphically. The value of the regression coefficient is an inference from sample data. The graph's horizontal axis shows possible levels for the unobservable

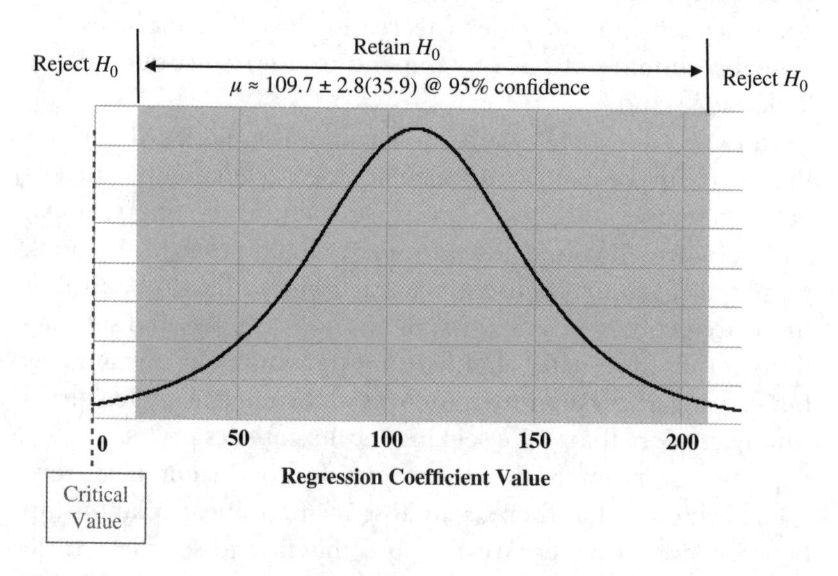

Figure 15.3 Determining Whether 0 Falls Within Confidence Interval

value of the mean slope for all homes in the population from which the sample was drawn, while the vertical axis shows the relative likelihood for the unobservable true value of the slope (μ).

The shaded area shows the confidence interval that is expected to capture μ with a 95% level of confidence. The light gray, dotted line shows the position of the critical value (CV) of 0. The CV falls outside of the confidence interval, so we have some comfort in rejecting the null hypothesis that there is an absence of a linear relationship between square feet and sales price.

Some students look at this graph and ask why the 95% confidence interval is ±2.8 standard errors instead of ±2. The reason goes back to what we discussed in Table 12.1 and Figure 12.2: we have a thin sample where $n = 6$ observations. Since n is fewer than 30, we resort to a t-distribution which has wider tails than a normal distribution.

We only have six observations to estimate two parameters, the y-intercept and the slope of the line. This estimation procedure chews up two "degrees of freedom," the number of observations available to form a statistical inference. The intuition is that we need two data points to draw a straight line, which leaves four remaining degrees of freedom from our sample of six observations.

In this case, just glance at Table 12.1: go down the column for $\alpha = 5\%$ until you get to five observations (the six observations available minus another one needed to estimate the intercept), and note that the table indicates that the desired confidence interval requires about 2.8 standard errors. Don't worry about the particulars – going into more detail is simply not worth the effort.

Other students often ask why we should reject the null hypothesis at the far-right part of the graph. If μ were equal to, say, \$220, wouldn't that add credence to our argument that the slope is not zero? I respond with agreement. Welcome to the messy language of numeracy.

The convention with regression software is to show p-values for a two-tailed test, where one may reject the null if the critical value is below or above the limits of the confidence interval. The reason is that two-tailed tests give higher *p*-values that are more conservative (that is, larger and less likely to fall below the agreed-upon 5% threshold for hypothesis testing). A numerate person would call this out when presenting conclusions to a boss or client.

The point to remember is that *p*-value collapses this complicated discussion about confidence intervals, critical values, degrees of freedom, sample size, standard errors, alpha, and so on into a single number. My experience is that audiences want a simple story that is accurate, if not complete. Just going to a regression output and pointing out that the *p*-value is below our rule of thumb of 5% is sufficient to keep a conversation going. Remember, our goal is simply to identify provisional patterns that may be tested rigorously later by propeller heads.

After asking whether the regression coefficient has the expected sign (it does) and is statistically significant (it appears to be), then we move on to the third step.

Here we address the idea of ***fit***, how well our model summarizes the data points. A high level of fit suggests that our model does a decent job of predicting or explaining things; a low level of fit suggests that we should not place too much credence in our model when making sense of the world around us.

We came across fit in our discussion of coefficient of determination in Figure 9.1, where we used the measure of r^2 to attempt to measure how much of the variance of one variable could be explained by a linear relationship with another. More importantly, we used the complement of this figure, $1 - r^2$, to show how much of the variance could not be explained by the relationship.

If the correlation coefficient were 1.0 or –1.0, then we would have a perfectly linear relationship between two variables, so the value for r^2 would be 100%, suggesting that movement in one variable perfectly explains movement in another. In my experience, this does not happen in the real world.

In the case of our home pricing model, we see near the bottom of Table 15.1 that the coefficient of determination (r^2) of our model is 70%, so its complement (1 – 70% = 30%) suggests that about a third of the movement in sales price cannot be explained by movement in square feet. Other factors likely contribute to home prices.

In regression analysis, the computer does this work for us and offers an additional summary statistic. Notice on the Excel output of Table 15.2 the label "Multiple *R*" and the corresponding value of 0.837. This is the same value of *r* shown in Table 15.1. The "*R* Square" label and related value of 70.0% shows the coefficient of determination between the independent and dependent variables.

Finally, note the "Adjusted *R* Square" label and the corresponding value of 62.5%. This is the model fit for the regression equation that includes both an intercept ($30,800) and a slope ($109.7 per square foot). This modified calculation of r^2 takes into consideration the number of observations and number of coefficients to be estimated (something we explore in the next chapter). The general rule is that adjusted r^2 delivers a lower percentage than does r^2 and thus represents a more conservative estimate of model fit.

Here, we have an adjusted r^2 of 62.5%, which says that 37.5% (say, three-eighths) of movement in home prices is not explained by our model. This is not a bad value of model fit. Since only about two-thirds of movement in home prices is explained by our model, we use caution when using our model to make predictions.

Putting this all together, we could say to our spouse:

Honey, I ran the numbers for recent home sales prices in our neighborhood. My analysis suggests our home might sell for about $442,000. I have some confidence in this estimate because my linear regression had a coefficient with the anticipated positive sign, the coefficient of $109.7 per square foot was statistically different from zero, and my model fit *suggested that only about a third of variation in sales prices is not explained by my model.*

Wow, what a mouthful. Did your little model consider that we remodeled the entire first floor last year?

Oops. She's got us. The fatal flaw with most models is the problem of omitted variables. Our model has just one independent variable. It's a tall order to make accurate predictions based on just one variable. We address this concern in the next chapter.

Before closing, please go back to Figure 15.1. Note how we drew a best-fit line through six observations over the domain from 1,800 to 4,100 square feet. Our sample has limitations for making predictions. Whenever using a model to make a prediction, consider where the value of the independent variable(s) used in the prediction falls in relation to observed data in the sample.

If we wished to use this model to make an estimate for a large house, say 6,000 square feet, we must call out that the *x*-value is beyond anything we've observed. The relationship between home size and sales price may not be linear, but a linear relationship could be useful to make predictions within a limited domain. We saw this happen in Figure 9.3.

We could also have a problem estimating the value of a small home with a square footage below 1,800. Our model has a y-intercept of $30,800, which is ridiculous. No one would pay anything for a home with zero square feet.

Our regression output alerts us to this limitation in the model. Note on Table 15.2 that the *p*-value for the intercept is 80.4%. This means that there is a very high probability that there is no significant difference between $30,800 and zero. We could have been justified in drawing our best-fit line through the graph's origin, but our doing so would have changed the slope of the line to a value other than $109.7 per square foot.

Just remember that all models are wrong, but some are useful.

Recap of Chapter 15 (Slope [$r(s_Y/s_X)$])

- We use models to explain the past and to predict the future.
- Correlation is the primary tool numerate people use to make predictions.
- Slope, scaling r by the ratio of standard deviations, allows us to predict quantities.
- Linear regression is a helpful tool in generalizing relationships from observed data.
- Coefficient sign, p-value, and adjusted r-squared permit evaluation of regressions.

16

CAUSATION

One theme of this book is that we study numeracy to better explain the past or predict the future. Only true nerds study this material because they find it inherently interesting. You're reading this book, I believe, because you seek to hone your ability to extract information embedded in unfamiliar data sets to make more informed contributions to organizations we serve.

The previous chapter focused on prediction. We showed how the concept of slope, as brought about by linear regression, permits predictions about unobserved phenomena. We don't need to know the causal mechanism between two variables to predict how change in one variable brings movement in another. We simply need the confidence to say that data in our sample is representative of what's going on in a larger, unobservable population.

Earlier my career, I set car insurance prices based on, among other things, the sex of the driver. Claims data I reviewed provided compelling evidence that men are more likely to have accidents than women. Why? I didn't know, I didn't care, and I didn't go there. The last thing I wanted to do was to get up in front of a group of employees, customers, regulators, journalists, or whoever and start a discussion about gender differences.

My professional concern was whether prices charged were sufficient to cover anticipated claims costs. Correlation analysis provided robust evidence to suggest that insuring men was more expensive than insuring women. I chose to charge males higher rates in markets for which I was responsible for pricing decisions and then move on to other issues. I enjoyed some professional success in my insurance pricing efforts, but I still cannot offer you an explanation of why men tend to drive cars into things more frequently.

The other purpose of numeracy is to put forward explanations. Explaining something introduces the idea of *Why?* into a discussion. When each of our children reached about four years of age, they started asking why things are the way they are. Offering predictions did little to satisfy their curiosity. The craft of explaining something requires tools beyond correlation.

An example of explanation without prediction is answering the question of why dinosaurs went extinct. These animals ruled the roost but then disappeared. Outside of science fiction, there is little to predict about future dinosaurs. Understanding why dinosaurs died out, however, is a worthwhile question as mankind ponders its future on this planet.

The reigning explanation is that a celestial object slammed into the Yucatán Peninsula about 66 million years ago. Debris thrown into the atmosphere blotted out the sun, killing plants dependent on sunlight. Dinosaurs and other critters down the food chain starved. Geological evidence found around the world is consistent with this story, so scientists have not been able to falsify this hypothesis.

A causal explanation is a story that seeks to show how *X* (in this case, a sizable object hitting the earth) brings about *Y* (the extinction of dinosaurs). Putting forward a causal explanation does more than satisfy curiosity. If we believe that a second event is a consequence of the first, then we may presume that efforts to modify the first variable may bring desired changes to the second variable. An understanding of ***causation*** allows us to help change the world in intended ways. This, in my opinion, is the ultimate source of power.

I am a devout capitalist, but I respect how much Karl Marx influenced global society. No study of the twentieth century is complete without consideration of how Marx's ideas changed countries and international relations. The guy punched way above his weight. An inscription on Marx's tombstone says "The philosophers have only interpreted the world in various ways. The point however is to change it." Ideas become significant when they effect change.

Figure 16.1 illustrates the influence of causal reasoning in a more modest domain. Before the pandemic, I exercised regularly at the gymnasium on our university's campus. On April 2, 2018, I came across the sign shown in the photograph. The photo is off-center because I did not want to get in trouble for taking a photograph in the men's locker room, so I just snapped one quick picture. Thankfully, no one noticed.

The sign says that the gym's water temperature is oscillating between hot, warm, cool, and cold temperatures, and the staff is trying to identify the cause. Fluctuating temperatures make for uncomfortable showers, as I soon learned.

The *Y* variable (volatility in water temperature) resulted from an unknown cause (the to-be-discovered *X* variable). If the staff could identify the cause and then make needed changes to the *X* variable, then the problem would go away. The staff must have done this because the water temperature was stable at my next visit, and the problem did not recur. The vignette of my calling the IT help desk in Chapter 13 is another example of searching for a causal relationship and then acting on provisional findings to bring about desired changed.

The question of how to determine if *X* causes *Y* has vexed philosophers for millennia. I offer a simple, three-step process that many find useful. One may argue that *X* causes *Y* by demonstrating that:

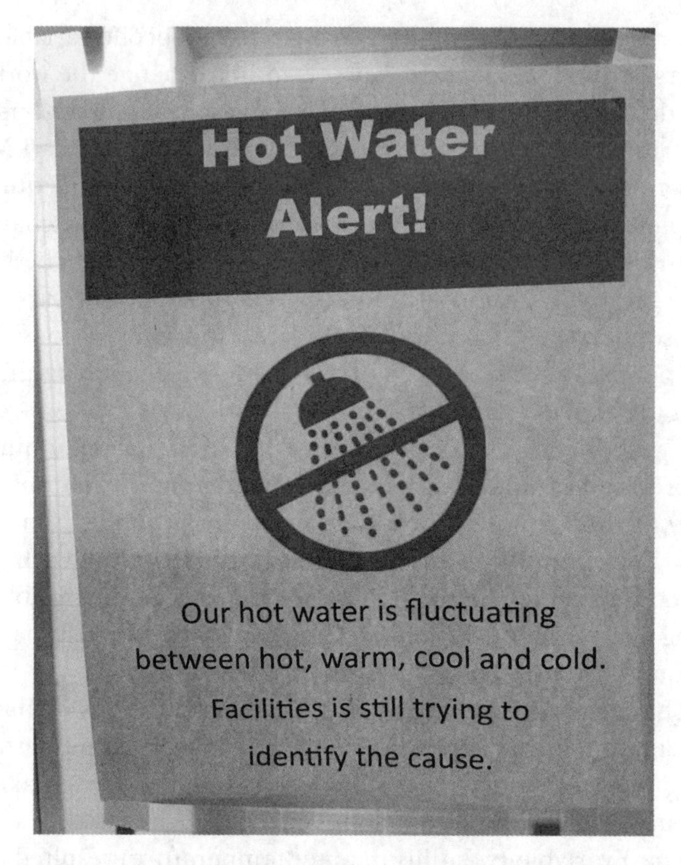

Figure 16.1 Photograph Taken at My Gym on April 2, 2018

1. *X* is correlated with *Y*;
2. *X* precedes *Y*; and
3. The usual suspects can be ruled out.

The first step is determining if there is correlation between *X* and *Y*. The use of *r*, discussed in Chapter 8, is a simple way to assess the strength and direction of a possible linear relationship between two variables. Using *r* works best if we have continuous, ratio data and if we assume that the relationship between two variables is linear. Even if we have doubts about using *r*, start with this tool and then, should early results look promising, call out this limitation and suggest that a propeller head could be brought in to kick the tires harder.

If r looks decent, spend another minute to calculate the p-value assessing whether r is meaningfully different from the critical value of 0.[1] If r is close to 0 and/or is based on a small sample, then we should be careful about arguing that two variables are correlated.

Please remember the warning you heard in school that correlation is not causation. The presence of correlation does not mean that one variable caused another. Among many possibilities, outside (lurking) variables may influence movement in both X and Y. We saw this in Chapter 8 when we noted correlation in our refrigerator's fruit drawer. Variation in apple and orange counts likely came from household grocery shopping and eating activities.

So, if we find evidence of correlation, we proceed to our second step of determining if X precedes Y. The fancy term is *temporal precedence.* It's hard to argue that X causes Y if Y happened first. A causal argument requires some type of time stamp to be associated with data points. In my field of accounting, we have time-stamped data on stock prices and information released to investors. It's relatively easy to argue temporal precedence when, say, a company announces previously unknown earnings data at a given time and then we see a stock price reaction following the news release.

Those who debated in high school have heard the Latin expression *post hoc, ergo propter hoc* (after this, therefore because of this). The term describes a logical fallacy concluding that because one event followed another, there is sufficient evidence to argue that the first event caused the second event.

---

[1]This is easy to do. All one needs is the value of r and the sample size (i.e., the number of pairs of data points). Here is one of many calculators available on the web to measure p-value for Pearson's r (retrieved 19 February 2021): https://www.socscistatistics.com/pvalues/pearsondistribution.aspx.

I demonstrate this problem when faced with a close Cleveland Browns football game. Too many times I watched my team snatch defeat from the jaws of victory. On one game, I turned off the television and later noted that the Browns won. I concluded that my not watching the final minutes of the game brought victory. This sounds absurd, but I now turn off the TV on close games to help my team prevail. Superstitions aside, temporal precedence, like correlation, is a necessary but not sufficient condition to argue causation.

The final part of this three-legged stool is ruling out alternative explanations. To argue that *X* causes *Y*, we must further show that other variables (e.g., *U*, *V*, or *W*) are not the reason that *Y* happened. An example was shown in my phone call to the IT help desk. I was asked to verify that my inability to connect with my university's website could not be explained by the absence of a Wi-Fi connection or by use of a specific browser.

Ruling out alternative explanations is the most difficult of the three tests for arguing causation. One could dream up an endless number of variables that could bring about a change in *Y*. We could say, for example, that invisible aliens from Mars tampered with the gym's water system. If we cannot use our senses to rule out the presence of the aliens, it becomes impossible to demonstrate conclusively that they were not the cause of oscillating water temperatures.

My standard for the third test comes from a line from the classic World War II film *Casablanca*. Humphrey Bogart shoots a Nazi to allow a couple to escape to neutral Portugal. Corrupt French policeman Captain Renault, sympathetic to Bogart's chivalry, deadpans to his subordinates that they should round up the usual suspects to investigate the crime and put the matter to rest. Renault simply must show that the crime was investigated before closing the file.

In this context, "usual suspects" is a metaphor for showing that other potential explanations for a causal relationship were considered but then rejected. We can offer evidence to show that these alternative causal factors do not explain the reason *Y* happened, but we will never have the time or budget to rule out an exhaustive list of alternatives.

Let's provide a stylized example to try to make these ideas clearer. Suppose our boss, the chief marketing officer (CMO) at our firm, seeks permission from a skeptical CFO to increase the advertising budget. The argument is that investing in a brand will elevate sales revenue for one of our products in a crowded market. The CFO had approved a test advertising campaign that began in January 2021. Now, with data available through March 2022, it's time to assess results.

Table 16.1 Influence of Lagged Advertising and Unit Price on Monthly Sales Volume

Month	*Unit Sales*	*Lagged Advertising*	*Unit Price*
Jan-21	3,050	–	6.25
Feb-21	3,158	100	6.25
Mar-21	3,300	250	6.25
Apr-21	3,650	500	6.25
May-21	3,900	750	6.25
Jun-21	4,000	800	6.25
Jul-21	4,200	850	6.25
Aug-21	4,200	900	6.50
Sep-21	4,300	975	6.50
Oct-21	4,400	1,000	6.50
Nov-21	4,500	1,000	6.50
Dec-21	4,133	500	6.50
Jan-22	4,002	250	6.50
Feb-22	3,975	100	6.50
Mar-22	3,701	50	6.50

SUMMARY OUTPUT

Regression Statistics	
Multiple *R*	94.2%
R Square	88.7%
Adjusted *R* Square	86.8%
Standard Error	162
Observations	15

	Coefficients	*p-value*
Intercept	(7,611)	0.420%
Lagged Advertising	0.827	0.001%
Unit Price	1,734	0.026%

We compile data shown in Table 16.1. There are four variables: the month associated with business activity, the unit sales of our product, advertising expenditures in thousands of dollars, and the price charged for our product in dollars per unit. Seeking to show whether there is a casual link between spending money on advertising and sales quantity, we make one adjustment and then draw on one more statistical technique.

The adjustment is to lag the advertising spending by one month. Our experiment started in January 2021 with a monthly budget of $100,000. We then raised the level of expenditure to $250,000 in February 2021.

A causal argument requires demonstrating temporal precedence, so we arbitrarily adjust the timing of the advertising expenditures by one month, matching the $100,000 January 2021 spend with February 2021 unit sales. February 2021 advertising of $250,000 is moved to be associated with March 2021 sales, and so on. The reasoning is that it takes time for advertising expenditures to influence customers to buy our product. The March 2022 advertising activity is not reflected in this sample.

The additional statistical technique is use of multiple linear regression, an expansion of the simple linear regression tool introduced in Chapter 15. Instead of assessing how a *Y* variable moves with variation in *X*, a single independent variable, this tool assesses how Y moves with variation in multiple independent variables (say, X and W).

Multiple regression analysis parses movement in Y among movements in independent variables. The regression coefficients show how much Y moves given a one-unit movement in X while holding W constant. The math associated with doing these calculations is beyond the scope of this book and need not worry us.

The dependent variable, Y, is monthly unit sales volume. The primary independent variable, X, is the level of advertising expenditures made in the previous month. We hope to show that increases in X bring about increases in Y. We add the price charged per unit as secondary independent variable, W.

We call W a ***control*** variable because we seek to rule out the argument that increases in units sold from January 2021 through March 2022 resulted from cutting prices to spur sales. Adding unit prices as a control variable allows us to show how sales volume is correlated with advertising after taking price changes into consideration.

At the bottom right side of Table 16.1 is simplified output from multiple regression analysis performed on Excel. I modified the formatting of the cells to reduce the number of significant digits and to express certain quantities as percentages.

Our boss, when shown the data presented in Table 16.1, looks like a deer in the headlights. The wall of numbers means nothing to him. We don't get snarky because he's got skills we'll never have. We say:

Okay, Boss, I have good news. What you're looking at suggests we have evidence that advertising increased sales volume during our trial period.

Great. Tell me more.

We're trying to sell the CFO that there is a causal relationship between advertising and increased sales. To show causation, we need to demonstrate three things: (1) increases in advertising are correlated with increases in sales, (2) advertising activity preceded sales increases, and (3) we may rule out the usual suspects of other possible influences.

Sounds good. Keep going.

The first test is correlation between advertising and sales volume. I ran a linear regression, and the output shows that for every increase in one thousand dollars of advertising, we saw an increase in 827 units sold. What's more, the "p-value" of 0.001% says that there is just a tiny chance that this relationship was no different from zero.

So, what you're saying is that the statistics show that increases in advertising bring increases in unit sales and this relationship does not look like a fluke?

Yep. We may tick off correlation on our checklist. Next, we can demonstrate "temporal precedence" by lagging the advertising spend one month. I ran the numbers so that, say, August 2021's sales volume was matched against July 2021's advertising efforts.

Okay, we're two for three. What do you mean by "usual suspects"? You seem a little straitlaced to appreciate Bryan Singer films.

Well, there you go again, underestimating me. But here we're talking about control variables used to rule out alternative explanations. I included monthly prices charged so that the advertising coefficient does not consider the influence of changing prices. The CFO would likely ask if sales went up because we cut prices rather than because the advertising worked. The price coefficient, which also appears to be statistically significant, is positive, which suggests that increased prices are associated with higher unit sales. Maybe our advertising increased customers' willingness to buy our product.

Got it. Your math says we're three for three and we can walk into the CFO's office with confidence?

Pretty much. You should understand that we have a thin sample with just 15 observations, so we're not going to win any math prizes with this data set. However, the adjusted R-squared of this model, which assesses how well it explains monthly movement in unit sales, is almost 87%. This means that only about 13% in monthly sales cannot be explained by advertising spend and price level. I'm willing to bet my career that our advertising worked.

This stylized exchange shows the rudiments of a causal argument. This is much more compelling than just measuring correlation or slope. The CFO will receive a cogent explanation for why we believe that advertising influences unit sales. If the causal argument is valid, then we have a basis for improving our business: controllable modifications to the *X* variable may be expected to bring desired changes in the *Y* variable.

The real world is rarely as simple as the example just given. There are many, many factors associated with phenomena in the world around us. It is relatively easy to measure correlation between two variables and then demonstrate that the relationship holds in new samples. It is an entirely different exercise to craft a causal explanation that is accepted by skeptical audiences.

Any review of global university rankings shows that so-called elite universities tend to be based in comparatively wealthy countries. Figure 16.2 offers various ways to try to explain this relationship. The four boxes represent constructs that I have not bothered to operationalize with variables. Let's just assume that we could define and measure the ideas shown in the boxes in a reliable and valid manner.

There are many plausible causal explanations for why there is an association between good universities and wealthy countries. Following are six of many possibilities. Technically, we're showing possible relationships among constructs instead of

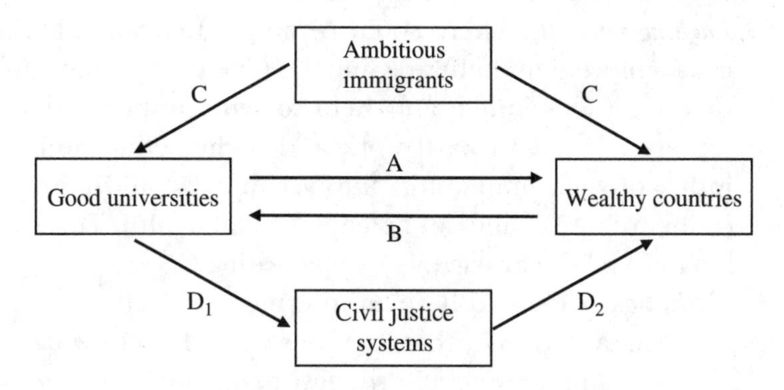

Figure 16.2 Correlation May Elicit Competing Causal Arguments

variables, so the proposed explanations would be propositions, not hypotheses:

1. ***Spurious correlation.*** It may be mere coincidence that good universities appear in wealthy countries. We mentioned this concept in Chapter 8. Spurious relationships are not expected to repeat out of sample, so any relationship seen today will not be expected to last.
2. ***Causation.*** There may be a simple causal relationship where graduates of high-quality universities (*X*) create wealth and bring about prosperous societies (*Y*). This story is represented by arrow A.
3. ***Reverse causation.*** We could have our story backwards, where our supposed independent variable is really the dependent variable: instead of *X* causing *Y*, what is really happening is *Y* causing *X*. Wealth comes first, and only wealthy countries have the resources to make the investments required to develop high-quality universities. This story is shown by arrow B.
4. ***Endogeneity.*** There could be a circular relationship between *X* and *Y* where they feed off each other. As countries get wealthier, they invest more in universities, and the rising quality of universities in turn produces more capable graduates who help create new wealth. This story is shown by the combination of arrows A and B.
5. ***Lurking variable.*** There could be an omitted variable that causes movement in both *X* and *Y*. Here, we could imagine that ambitious immigrants help societies improve. These settlers value both hard work and higher education. An influx of such immigrants into a country could raise the caliber of universities and create societal wealth. This story is shown by the two arrows labeled with a C.
6. ***Mediation.*** We could have an intermediate factor (say, *M*) between *X* and *Y* so that *X* causes *M* and then *M* causes *Y*. For example, we could suggest that some graduates of high-quality universities pursue legal careers and that the

presence of well-trained lawyers brings effective civil justice systems that set the stage for private-sector wealth creation. This story is shown by the arrows D_1 and D_2.

The search for causal explanations may become even more complicated. We could consider the influence of multiple lurking variables on the *X* and *Y* variables. We could also have several mediating variables working in parallel or in series between *X* and *Y*.

An example of multiple mediating factors is given in the legal decision *Palsgraf v. Long Island Railroad* (248 N.Y. 339, 1928), discussed in many introductory tort classes. The approximate facts are that an employee helped rush a passenger on to a departing train, which led to someone dropping a package containing fireworks, which led to an explosion as the package hit the ground, which caused a scale to topple and hit a bystander, who subsequently developed a stammer. The railroad was not found negligent because this bizarre sequence of events was not deemed to be foreseeable by the railroad employee.

We could have ***moderation***, not shown in Figure 16.2, where the presence of another variable causes the *X* variable to have an outsized influence on the *Y* variable. An example is adding chocolate syrup to vanilla ice cream. Eating ice cream (the *X* variable) brings some people a pleasant snacking experience (*Y*). Adding chocolate syrup amplifies the pleasure of the snacking experience. The combination of chocolate and vanilla provides an interaction that would probably not result from using, say, ketchup.

The point of all these examples is to highlight the vast number of possible explanations to account for correlation between two variables. It can take researchers a long time to falsify complicated causal explanations. I met an economist who spent a career trying to explain why Silicon Valley flourished. Finding a causal explanation would be incredibly helpful for national governments seeking to set industrial policy. Unfortunately, the economist was never able to effectively rule out competing explanations to identify a single answer.

When trying to make models of the world around us, we must assess how many variables to consider. Where we draw the boundaries of a model requires judgment. There are so many connections in the world around us – remember the so-called butterfly effect, where flapping wings may lead to extreme changes in distant weather patterns – that we could include dozens or hundreds of variables to sort through causal explanations.

Some people turn to the idea of ***Occam's razor***, where, among competing hypotheses of equal effectiveness, one should select the one with the fewest assumptions. An accepted view of numerate people is that simpler models are more understandable and are preferred. No one expects us to create a theory of everything.

In classroom discussions on model creation, I'm sometimes asked how many variables should be included in a model. My response is that it depends on whether the model is to be used to predict or explain something.

If the model is to be used to predict the future, then add independent variables if doing so provides significant improvement to model fit, as reflected in measures such as adjusted *R*-squared in linear regression output. Piling on independent variables brings diminishing returns. My experience is that two or three predictors often gets us to a good place.

If the purpose of the model is to explain the past with a causal argument, then add as many control variables as needed to rule out the usual suspects. If we don't do this, our clients or bosses will say that the model is incomplete because we neglected to consider the influence of their pet explanation.

If you must pile on control variables to put someone else at ease, do so and then point out that some of the independent variables will almost certainly be correlated, which complicates efforts to make accurate predictions. If the boss or client also wants a prediction, then show how some control variables may be safely removed from the model without harming model fit.

Propeller heads have fancy ways to sort through such multicollinearity and endogeneity.

The big idea from all this detail is that correlation is measured while causation is argued. A software package may return a precise measure for correlation, but no computer program is yet able to argue causation. While a human being may be able to put forward a persuasive causal argument, they will never be able to prove causation because there is an infinite number of alternative explanations to be ruled out. The real test is whether you would bet your career on a proposed causal explanation when forced to make a risky decision.

A parting thought is that causal explanations need not be complete to be useful. The English physician Edward Jenner (1749–1823) noted that milkmaids were immune to smallpox and reasoned that exposing patients to a small quantity of pus from a diseased cow brought immunity. Subsequent development of the field of immunology has led to the saving of millions of lives. Jenner did a lot of good without offering a complete explanation for why his procedure worked.

Recap of Chapter 16 (Causation)

- We may use numeracy to move beyond prediction to suggest causal explanations.
- Good causal explanations allow us to bring desired change.
- Causal arguments require showing correlation, temporal precedence, and ruled-out alternatives.
- It is possible to suggest competing causal arguments.
- It may be expensive – or impossible – to falsify some causal arguments.

17

SCIENCE

This brief chapter organizes our tools in a manner designed to make them more useful. Figures 17.1 and 17.2 summarize ideas that have taken me decades to learn. These diagrams show how numeracy integrates two types of reasoning.

Deduction, promoted by Aristotle and his followers, means applying general principles to specific circumstances. An oft-cited example is the following syllogism, a collection of three statements composed of two premises and a conclusion. If we believe that (1) all men are mortal and that (2) Socrates is a man, then we may conclude that (3) Socrates will die. The historical record supports this prediction.

Deduction combines premises (which may be assumptions and/or previously demonstrated generalizations) to develop new insights. Consider how deduction is used in Euclid's *Elements,* an introduction to geometry that builds an impressive collection of ideas from just five postulates that are assumed to be true.

If we assume that a straight line may be drawn between any two points and a circle may be drawn from any one point, then we may use deductive reasoning to demonstrate how drawing circles of equal radii from two points creates an equilateral triangle. This resulting proposition may then be used to demonstrate how to bisect an angle. This second proposition in turn may be used to demonstrate more elegant downstream statements. In other words, deduction uses previously demonstrated relationships as scaffolding to build ever-more sophisticated conclusions.

Perhaps the ultimate accomplishment from deductive reasoning is Andrew Wiles's proof for Fermat's Last Theorem, which states that no three positive integers a, b, and c satisfy the relationship $a^n + b^n = c^n$ for any positive integer n greater than 2. The French mathematician Pierre de Fermat claimed in 1637 that he was able to prove this simple statement without showing his work. Wiles used building blocks developed by predecessors over three centuries to develop a proof in 1995. Wiles, like Newton, saw further because he could stand on the shoulders of giants.

The validity of conclusions reached depends on the quality of the premises and the rigor with which they are combined. Flawed premises may lead us astray. The American satirist Ambrose Bierce (1842–ca. 1914) used this example to poke fun at deductive reasoning.

> The basis of logic is the syllogism, consisting of a major and minor premise and a conclusion:
>
> 1. Major premise: Sixty men can do a piece of work 60 times as quickly as one man.
> 2. Minor premise: One man can dig a post-hole in 60 seconds.
> 3. Conclusion: Sixty men can dig a post-hole in one second.
>
> This may be called the syllogism arithmetical, in which, by combining logic with mathematics, we obtain a double certainty and are twice blessed. (Bierce, 1911)

Bierce shows how conclusions reached from deductive reasoning are only as good as the starting premises. It is absurd to believe that enlisting five dozen people to do the job of one will always reduce a task's cycle time by a factor of 60.

Shaky premises bring wobbly conclusions. Euclid's fifth postulate says that only one set of parallel lines may be drawn through two points. Later mathematicians showed that space may be constructed so that lines drawn through two points must intersect (as on the surface of a sphere) or that many lines drawn through two points will never intersect (as on the surface of a horse saddle). Modifying one postulate changes the conclusions that may be reached from deductive reasoning. Predecessors' shoulders are not necessarily steady.

While deductive reasoning is a helpful means of taming an uncertain world, we must add a second form of reasoning to our toolkit. ***Induction*** abstracts from specific examples to suggest general principles that extend beyond the domains of the observed data.

A toothpaste manufacturer seeks to boost sales by securing endorsements from health-care providers. The company surveys 100 dentists and finds that 80 would recommend the toothpaste in question. The firm may claim that four out of five of all dentists recommend their brand to patients. Induction means that the generalization applies to dentists outside of the sample, though with some margin of error.

The early-modern philosopher Francis Bacon (1561–1626) receives credit for suggesting that one may observe phenomena in the world around us and infer broad statements that describe the unobserved data beyond the domain of the observed facts. Inferences are informed guesses about relationships outside the sample.

The problem with induction is that conclusions reached may not apply to all future circumstances. An observer may experience many repetitions and still fail to show that the relationship is always and everywhere true.

Bertrand Russell (1872–1970) described the problem of the chicken. Every morning the sun rises and the farmer delivers a meal. Another day, another sunrise, another meal. The chicken may justifiably believe that meals necessarily follow sunrises. Experience shows that this relationship holds – until it doesn't. One day, the farmer arrives without breakfast and wrings the chicken's neck (Russell, 1912).

As discussed in Chapter 13, Karl Popper (1959) pushed this idea forward with the metaphor of white swans. Suppose someone notices that every observed swan is white. While they accumulate overwhelming amounts of corroborating evidence to support the inferences that all swans are white, there is no basis for ruling out the possibility that a black swan lurks around the corner. If, however, a distant observer in Australia finds a single black swan, then they disprove the generalization. A statement based on induction may never be proven true but may be shown to be false with the presence of a disconfirming observation.

Thus, deduction offers the prospect of proof, but we need airtight premises to craft valid arguments. Induction is grounded in real-world observations, but we struggle to convince ourselves that inferred relationships are always and everywhere true.

The nineteenth-century American philosopher Charles Sanders Peirce (1839–1914) suggested a third approach that blends induction and deduction. This hybrid, called ***abduction***, puts forth inferences as tentative suggestions about how things in the world around us are organized. Each guess is tested when it is applied to a domain outside of the original data set.

If the guess successfully predicts or explains things in the new sample, then the inference gains credence as a useful tool to make sense of a messy world. If the guess fails to offer a satisfactory prediction or explanation, then the guess is knocked down a peg or two and becomes susceptible to modification or replacement.

This process of trial and error continues as the generalization is modified to reflect an ever-expanding set of data against which the guess was tested. At some point, people declare victory and consider the generalization to be useful, even if it cannot be proven to be true.

After a sufficiently large number of trials, chemists agree that the element aluminum melts at about 660° C. We cannot prove that this melting point applies always and everywhere for all aluminum in the universe, but we have enough evidence to suggest that we may rely on this generalization when working with metals here on Earth.

Modified inferences – those subjected to repeated tests – are never proven to be true. They are simply maintained until they are no longer useful. This approach of dispensing with a search for truth and instead simply asking whether a generalization works is known as pragmatism, a school of thought suggested by Peirce and his followers (Menand, 2001). Pragmatism is perhaps the only "ism" in philosophy for which Americans may take credit.

Figure 17.1 shows how abduction blends inductive and deductive reasoning in an endless fashion to refine generalizations and make them more useful. Pragmatists leave searches for truth to philosophers.

An example of abduction is the approach astronomers have taken to refine our understanding of the motion of planets. Ancient observers noted how planets wander across the sky while

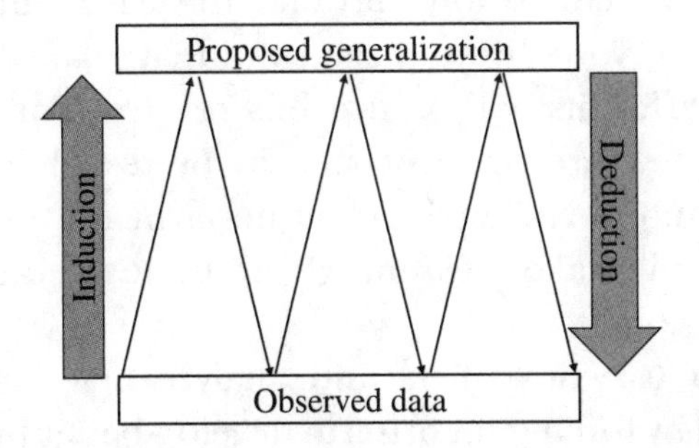

Figure 17.1 Science Blends Inductive and Deductive Reasoning

stars don't move. Ptolemy reasoned that the planets and sun revolve around Earth. Copernicus applied this generalization to a wider body of observations and suggested that Earth is just another planet and that they all revolve around the sun in circular orbits. This second explanation accounted for celestial movements with fewer assumptions, an example of applying Occam's razor, discussed in the previous chapter.

More sophisticated measurements of planetary positions, notably by the Danish astronomer Tycho Brahe, led German mathematician Johannes Kepler to suggest that planets follow elliptical orbits, where the sun is located at a focus within an ellipse.

The elliptical model did not perfectly predict future locations of orbital bodies. A more robust explanation, showing how the farthest point from the focus rotates around the central body, can better predict a planet or satellite's position. Einstein's general theory of relativity may be used to explain this precession.

I believe that general relativity will not have the last word on the movement of celestial bodies. Future generations of astronomers will develop even more refined explanations of how objects move through the heavens. Yet, we may still rely on general relativity to help us use global positioning satellites to monitor our location when we travel.

This example shows an iterative relationship between induction and deduction. Any inductive exercise (e.g., suggesting that planets and the sun revolve around the earth, the earth and other planets revolve around the sun in circular orbits, planets have elliptical orbits, elliptical orbits revolve around a central star) brings new predictions that may be tested against bigger data sets. Edmond Halley, who had never heard of general relativity, could still make a stunningly accurate prediction for the return of his comet.

Those of us who collect and analyze new data serve the greater good by forcing theorists to develop better predictions or explanations. The great astronomer Vera Rubin, who noted that distant stars orbit galactic centers faster than expected and thus provided evidence of dark matter, viewed herself as an observer whose role was to confound the theorists and spur them to create better models (Hagerty, 2016). The process stops when people are satisfied with the quality of the predictions or explanations put forth by the most recent generalization.

To use a phrase offered by a civil servant with whom I had developed a cordial relationship, we keep going until we find a solution that consistently yields results that are close enough for government work. However, we note that we haven't proven anything, and we accept that our framework may have to be discarded if our environment changes. We retain imperfect theories until we find something better. Economist Paul Samuelson (1915–2009) noted that it takes a theory to kill a theory because anomalous facts can only dent a theorist's hide.

Figure 17.2 ties things together, offering an overview of how we use the tools presented in this book to follow an abductive reasoning process.

Numeracy begins in the bottom box labeled Observation. We look at data in the world around us and recognize that what we see is likely just a sample of a larger phenomenon that we seek to better understand.

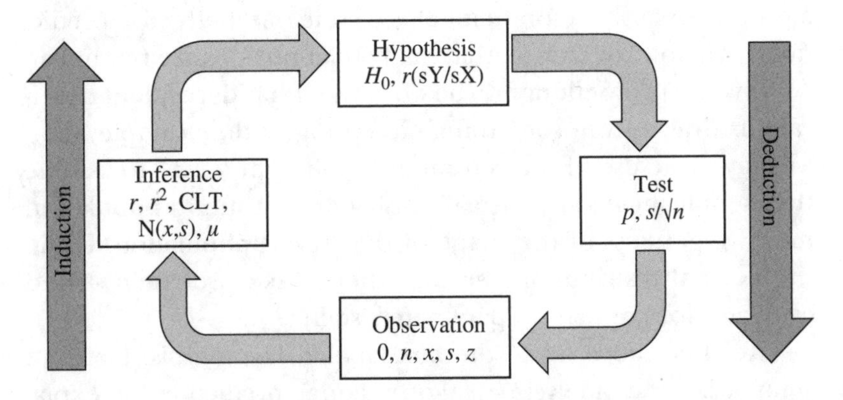

Figure 17.2 Numeracy Integrates Concepts Covered in This Book

Drawing on our understanding of 0, we ask if we may add, subtract, multiply, or divide our observations and whether they may be measured with increasing levels of granularity. Results of this analysis indicate whether we're dealing with dichotomous, discrete, or continuous data and whether they should be classified as nominal, ordinal, interval, or ratio scales. If what we have is something other than continuous, ratio data, we then share this limitation with our client or boss.

We then call on our three familiar friends, n, x, and s, to assess the robustness of our sample, measure central tendency to enforce comparisons, and consider the variance of observations around their mean. Reflecting on relationships among these three sample statistics offers clues about how dangerous it is to use our observations to predict or explain things. We also use z to look for outliers, which reveal information or suggest the presence of measurement problems. Of course, we may choose to focus on other measures beyond x and s in our analysis.

We move on to the box labeled Inference. Here we look for the possibility of broader relationships that extend beyond the sample at hand. The core tool used to form predictions is r, which assesses the strength and direction of a linear relationship. From r, we may calculate $1 - r^2$ to show how much movement in one variable may not be used to explain movement in another – a process of falsification that helps us rule out ineffective predictions. We note, of course, that many relationships are not linear.

If we're lucky enough to be working with independent observations drawn from something close to a random sample, then we may make use of the central limit theorem (CLT) to assume that sample means are normally distributed around a population mean, regardless of the shape of the parent population. Using the normal distribution, we may then make probability-based estimates for population parameters such as μ.

We then move up to the box labeled Hypothesis. Here we summarize our analysis in a provisional prediction or explanation. To mitigate the risk of confirmation bias, we express hypotheses in null form. If we choose to express our hypothesis as a linear relationship between two variables, we use a measure of slope to assess how much movement in one variable brings corresponding movement in our variable of interest.

We then move to the box labeled Test. Here we see how well our hypothesis stands up to scrutiny when we apply the provisional relationship to a new sample of observations. We use p-value to assess the likelihood that an observed value (such as a regression coefficient) is more than a desired number of standard errors removed from a baseline measure (such as zero). A low p-value offers support for rejecting the null hypothesis and concluding that something interesting is afoot.

Deciding whether to retain or reject the null hypothesis is our choice, regardless of the p-value resulting from our test. We recognize that any conclusion we reach is provisional. If we have the time and budget, we should keep testing a surviving hypothesis to see if it is effective when applied to new data sets. If you're uncomfortable with all the assumptions enumerated in this chapter, I take consolation that you paid attention.

Recap of Chapter 17 (Science)

- Numeracy combines inductive and deductive reasoning.
- Tools taught in this book help us create and test hypotheses.
- There is no statistics police that force us to reject or retain a hypothesis.
- If we have the time and budget, we seek to test surviving hypotheses on new data sets.

18

QUESTIONS

To wrap up this book, I return to Oliver Cromwell's warning that we could be wrong. The tools discussed in previous chapters serve our efforts to make predictions or explanations in the messy world around us. Numerate people have helped us eradicate smallpox and land rovers on Mars. Becoming numerate, however, should not give us a false sense of security that we are masters of the universe.

The scope of what we do not understand overwhelms what we do. We are the blind men trying to make sense of the elephant. In our search for bigger pictures, we must always note that our conclusions are only as good as our data and assumptions.

To help keep us grounded, I provide a dozen questions to ask when offered new data or analysis. The purpose of these questions is not to obtain crisp answers. Instead, these questions are meant to spur dialogue and spark ideas for how the analysis may be refined.

1. *What's the goal?* Always ask what problem we're trying to solve before looking at data or computer output. When

handed a spreadsheet, it is so tempting to let one's eyes scroll to the farthest right column and then drop to the bottom row to search for a summary figure. I guarantee that you do this when you receive performance reports for your retirement account. The trouble is that homing in on an available "answer" influences how we frame questions. A more disciplined approach would be to begin with an open-ended statement like *So, why are we talking about this?*

Another temptation is to start pushing buttons on a computer once someone shares a new data set. I feel an overwhelming need to calculate summary statistics and run some preliminary tests before sorting through what we're trying to accomplish. I have learned the hard way that doing so creates pages of statistical output that bring little more than fatigue and confusion. Haphazard button pushing has caused me to start crafting stories based on numbers that show up early on. Such analysis turns on study of a particular tree rather than on evaluation of the forest.

The goal of this book is to help you use some simple tools to better explain the past or predict the future. Half the journey is simply deciding on where you're going. Once you get clarity on the goal, write it down, share it, and solicit feedback. You will be amazed at the scope of argument that emerges before you start talking about any numbers.

Sorting through this debate brings an enormous reward. If we obtain agreement on what we seek to do, then we are much more likely to answer questions that interest our bosses and clients. The math required to accomplish the desired task is often incidental to framing the problem. Asking good questions better positions us to reveal information, something Claude Shannon described as a surprise.

2. *How did we get these data?* Another theme of this book is that bad samples give rise to bad inferences. As we learned from the *Literary Digest* story, large sample sizes are no protection against making catastrophic mistakes. We also learned that the dots we collect influence the dots we connect. The act of data collection begins to limit our ability to think broadly. Every data collection effort brings its own baggage. When approaching a new data set, reflect on its limitations. Meaningful data sets you touch will be samples of broader, unmeasurable populations. I doubt that there is any such thing as a completely unbiased sample.

   In my field of accounting, financial statements reflect samples of unimaginably large populations of business transactions. Retained earnings attempts and fails to cumulate business activity since the inception of the firm, investment accounts attempt to reflect estimates of future transactions over an indefinite time horizon, and expense accounts seek to record activities of suppliers, competitors, customers, tax authorities, and regulators who operate at the amorphous boundaries of the firm.

   Limitations in accounting systems constrain our ability to capture the results of all transactions. Accounting information never has and never will emerge from unbiased samples.

   To do our best work, we need to consider the inevitable biases that arose as we obtained the data. Ask yourself and others why the data are not representative of the population plus what kinds of observations likely went unobserved. Answers to these questions will give qualitative guidance to our clients and bosses as we discuss limitations of our analysis.
3. *Do observations go beyond continuous, ratio data?* The reasoning covered in this book rests on the use of x, s, and r. Calculating these measures requires subtraction and division among observations. In Chapter 2 we saw that these

operations may be problematic when we're faced with dichotomous and discrete data or data arranged in nominal, ordinal, or interval scales. Only continuous, ratio data may be safely subtracted and divided.

When we come across diverse data sets, we should do the best analysis we can with the tools at our disposal. If we find a surprising relationship, then we should highlight the limitations of our analysis and ask for permission to bring in a propeller head to kick the tires.

Should we obtain approval to spend the time and money to bring in a professional statistician, then we should do three things. First, make clear to our boss or client that fancy math will not provide a definitive answer; instead, manage expectations by sharing that sophisticated tools will simply help us further reduce uncertainty. Second, share the story behind the story to the statistician so that they may use their judgment to frame the analysis. Finally, give the statistician a finite time or monetary budget. I have yet to work for an organization that says that we have all the time and money in the world.

4. *Are the observations interdependent?* Traditional hypothesis testing assumes that sample means are normally distributed. The 68/95/99.7% rule simplifies calculations of confidence intervals and *p*-values. Unfortunately, many samples come from living, breathing systems chock-full of feedback loops. The math we've studied assumes that observations reflect independent events such as flipping coins, rolling dice, or dealing cards.

   Human behavior is fraught with interdependence. Investors absolutely watch each other as they trade securities. Doing so brings the risk of contagion, where extreme outcomes occur more frequently than would be expected assuming normal distributions.

   On December 19, 1973, TV talk show host Johnny Carson made a brief joke about an alleged toilet paper shortage.

Viewers paid attention and stocked up. Others took note and followed suit. Soon, store shelves were empty. Toilet paper purchasing volume spiked to levels that were unforeseeable to marketers and manufacturers whose job it was to match supply with demand. Their decision rules did not take into account the possibility of interdependent buying behavior.

If you believe that observations are subject to interdependence, warn your boss or client that they should devote time to preparing contingency plans to handle extreme events.

5. *How big is the sample?* Reflect on the number of observations (n) when faced with a new data set. I argue that there is a Goldilocks range for an appropriate sample size. As we've seen, thin samples bring the risk of bad inferences even if our math is perfect. Very small samples are associated with wide confidence intervals that do little to reduce uncertainty. Basing inferences on very thin samples is like chasing butterflies with ridiculously large nets.

   Yet, we should also be wary when we have very large samples. The first problem is that calculations for standard error ($s/\sqrt{n}$) become quite small when n is huge, making the two-standard error test for statistical significance a low hurdle. Second, there is always the risk of aggregating dissimilar observations when we have data collected over long periods of time or across large swaths of geography.
6. *What information was lost in aggregating the data?* Remember that numbers in a data set are crude attempts to signify messy phenomena in the world around us. Any effort to reduce people, places, or things to summary measures brings loss of information. Reflect on all the context that was lost when we studied winning times over the years at the men's Olympic 100-meter dash in Chapter 3. No two races were identical, so we should be cautious when treating observations as if they were fungible.

When looking at aggregated data, ask how the individual observations differ from each other and then what problems result from their heterogeneity. In accounting, much archival research is devoted to panel data composed of variables of interest (e.g., revenue and net income) gathered across companies (e.g., Daimler, Ford, General Motors, Honda, Toyota, Volkswagen) and over time (1980 through 2020). Among the many problems is that no two companies recognize revenue or calculate net income in identical ways and that the accounting rules used to make these calculations changed over time.

7. *How valid are the variables?* As discussed in Chapter 13, variables attempt to define and measure elusive constructs used to develop theories. If we are attempting to predict future home sale prices based on home size, we need to find a way to measure price and size. If what we measure does not capture the idea behind our construct, then our work won't be particularly useful.

   How to define value has vexed economists for centuries. The obvious solution – looking up recent sales price documented in the local county recorder's office – is not the definitive answer.

   Among many issues to consider are the recency of the sale (real estate markets change, so a sale a decade ago may not be informative today), uncaptured side deals associated with the sale (the seller throws in the appliances and agrees to pay to have a treehouse removed), and the inability of the sales price to reflect cost of replacement when the new homeowner needs to insure the dwelling.

8. *How reliable are the variables?* Should we get comfortable with validity, we must still worry whether different people will measure a variable in the same way. If we want to measure house size, how do we know that real estate authorities are consistent in measuring the number of square feet in a partially finished basement, in a loft that has a severely

slanted roof, or a putative three-season room adjacent to the kitchen?

Even if the measurement technique is reliable, there's no guarantee that the equipment used is up to the task. I am sure you have used a bad scale to weigh yourself at some point. You step on the scale and get a surprising reading. You step off to let the scale reset. You step on a second time and get a different reading. You repeat the process and get a third reading. Imagine how often this problem plays out in organizations that are in the business of measuring things.

9. *What goes wrong if we have a false positive?* We use *p*-values to assess whether something interesting is afoot. The logic of hypothesis testing involves determining whether there is no significant difference between a variable of interest and a critical value, such as zero, to suggest that nothing interesting is happening. (Yes, I know this use of a double negative is confusing.)

   A *p*-value of 5% suggests (assuming we may rely on our sampling methodology and the normal distribution) that there is a 1-in-20 chance that nothing is afoot. Yet, 5% can be a long way from zero. There is always a possibility that our test suggests that something interesting is happening when this is not the case.

   Being numerate means thinking ahead about the consequences of our statistical tests showing a false positive. If the consequences are grave, then we should be hesitant to take action when faced with evidence of statistical significance.

   On January 25, 1995, a team of Norwegian and American scientists launched a civilian rocket off the northwest coast of Norway. Russian sensors detected the event and noted that the trajectory resembled what a U.S. nuclear-armed submarine-launched missile would follow. Russian president Boris Yeltsin had to decide whether to launch a retaliatory strike against the United States.[1] Cool heads prevailed, and the world avoided a catastrophe, showing that Russian leaders understood the risks and consequences of Type I errors. We're here today because Russia has great math teachers.

10. *How should we think about risks of false negatives?* If we decide that we really don't want to have false positives, then we create the corresponding problem of producing a greater likelihood of false negatives. Type II errors mean that our statistical tests fail to recognize that something is really going on.

    An example would be calibrating a fire alarm so that it ignores everyday occurrences like burned toast. The adjustment raises the density of smoke required to sound an alarm. Should there be a real fire, valuable time could be lost if the reset sensor failed to notify members of the household that a fire emerged.

    It is difficult to reduce the likelihood of both false positives and negatives at the same time. In the U.S. criminal justice system, we have made the decision to calibrate criminal justice procedures so that we would rather have a guilty person go free (a Type II error) than convict an innocent person (a Type I error). We still get it wrong, as evidenced by news stories offering examples of both problems.

11. *What are the boundaries of our prediction or explanation?* Our ability to predict or explain is constrained by limitations in our sample, variables, and tools. The world out there is an incredibly complicated place. We should always ask whether findings identified in one domain (say, higher education in the United States today) apply to another domain (say, secondary education in Egypt a generation later).

---

[1]"Norwegian rocket incident," Wikipedia, 17 Dec. 2020. Web. 5 Mar. 2021.

12. *How do we summarize all of this work on a single sheet of paper?* I have learned the hard way that our work is meaningless if no one pays attention to what we find. Every boss, client, or colleague is both tired and distracted. Influencing others requires the ability to boil down thinking, research, data collection, and statistical testing to something approximating a 30-second television commercial.
    Chances of being an influencer skyrocket if one can package the thesis, relevant evidence, and story onto a single sheet of paper showing, at most, three significant digits. Success is hearing the magic words, *Tell me more.* We may then pull out a thick pile of data and analysis to give credibility to our summary message.

We end our journey where we began, with the story of Edmond Halley, who astounded his followers by accurately predicting that his comet would return in 1758. Halley sorted through the issues enumerated in this chapter to deliver a strikingly simple, understandable, and accurate prediction. Let us aspire to meet this standard.

Recap of Chapter 18 (Questions)

- Numeracy sparks discussion rather than solves problems.
- Numerate people ask questions about goals, assumptions, and data.
- We remember that all we're trying to do is extract information to reduce uncertainty.
- We make a contribution by helping colleagues put forth better predictions or explanations.

APPENDIX A – GLOSSARY

Becoming numerate requires developing a vocabulary. Following are my definitions for certain, specialized words numerate people need to know. I emphasize that the definitions express what these words mean to me. These words are set in bold italics in the book when they are introduced.

Abduction. A form of reasoning that seeks to find the simplest statement that makes explanations or predictions from the data at hand. I define this term to mean an iterative process of testing a hypothesis against new data sets and then making informed modifications to the hypothesis as experience accumulates from successive tests. We stop when our model is sufficiently useful.

Alpha [α]. A way of showing how worried we are that predictions will be wrong over multiple trials. If $\alpha = 5\%$, then we expect our statistical tools to help us create predictions that prove to be correct in 19 out of 20 trials.

Arithmetic mean [x]. A measure of central tendency calculated by dividing the sum of observations by the number of observations. I view the mean as the place where a fulcrum would balance a series of observations laid out as identical coins at appropriate intervals on a flat board. The arithmetic mean tends to be influenced by outliers.

Average. A measure of central tendency for observations within a group. Averages facilitate comparisons within or across groups. There are many measures for average, of which arithmetic mean ($\bar{x}$) is the best-known.

Bin. The use of buckets to collapse continuous data into discrete ranges so that observations may be counted and displayed in a histogram.

Causation. A story suggesting why one variable (X) brings about change in another variable (Y). Correlation is about what; causation is about why. In this book, we say that causation may be argued if we can show that X is correlated with Y, X precedes Y, and other variables (say, U, V, and W) don't matter.

Central limit theorem (CLT). A statistical finding where the distribution of arithmetic means from randomly drawn samples approaches a bell curve as the number of samples increases, regardless of the distribution of the population. Put simply, random samples may be used to make inferences about many kinds of populations. Using the CLT requires that the observations we analyze are independent and are selected in an unbiased manner – two strong assumptions.

Chebyshev's theorem. For any distribution, the minimum fraction of observations that must fall within a given number of standard deviations from the mean. Just remember that 75% of all observations must fall within two standard deviations of their arithmetic mean, regardless of the shape of the distribution.

Coefficient of determination [r^2]. The square of the correlation coefficient, providing a value between 0 and +1, which assesses how well movement in one variable may be used to assess movement in another, assuming a linear relationship between the two. The value of $1 - r^2$ assesses the degree to which movement in one variable does not explain movement in another variable through a linear relationship.

Coefficient of variation [s/x]. A measure of variance calculated by dividing a sample's standard deviation by its arithmetic mean. This ratio offers a unitless measure that may be compared across data sets. A value over 100% suggests it is dangerous to simply use x to explain or predict something.
Confidence level. The likelihood that we believe that our statistical tools will lead us to a prediction that will ultimately prove to be correct. Expressed as a fraction, confidence level is the complement of alpha (100% = confidence level + α). If $\alpha = 5\%$, then our confidence level is 95%, and we expect that our predictions will be accurate in 19 out of 20 trials.
Confirmation bias. The tendency to interpret data in a way that corroborates preexisting beliefs. Failure to mitigate confirmation bias risks scientific progress because we may hold on to hypotheses that are wrong. One theme of this book is to pay special attention to outlier observations that do not fit our tidy models for how the world works.
Construct. An idea used in theories that is difficult to define and measure. Intelligence and job performance are examples of constructs.
Continuous. A type of variable that may be expressed in increasingly granular manner. Examples are height, weight, temperature, and accounting balances, which may be measured to sizable numbers of decimal places. Most of the statistical tools discussed in this book are designed to be used for continuous variables.
Control. A variable included in a regression to help rule out the influence of the "usual suspects" of variables beyond the one that interests us most when examining changes in our dependent variable. Including controls in regressions allows us to provide evidence that we considered the third condition needed to argue causation.
Correlation. A situation where movement in one variable is associated with movement in another. Correlation is the basis for prediction.

Correlation coefficient [*r*]. A measure of the strength and direction of a linear relationship between two variables. Values range from –1 to +1.
Craft. A discipline that blends science (where there are agreed-upon answers) and art (where there are many defensible points of view). Numeracy is a craft.
Critical value. The benchmark for a test of statistical significance. A typical statistical test assesses the likelihood that a particular value of a variable is significantly different from a critical value.
Deduction. A form of reasoning that moves from general principles to specific circumstances. If we believe that all men are mortal and that Socrates is a man, then we may deduce that Socrates will die.
Descriptive statistics. A collection of measures used to find quantitative ways to summarize data sets. Commonly used measures are n, x, s, z, r, and r^2.
Dichotomous. A type of variable that may have just two values (e.g., heads or tails, on or off, pass or fail). Ask for help if you must analyze data sets in which the dependent variable is dichotomous.
Discrete. A type of variable that may take on only a certain number of values, like a rolled die or dealt card. Ask for help if your dependent variable is discrete and be cautious when using sample statistics to summarize discrete data sets.
Empirical rule. A handy trait of normal distributions where about 68%, 95%, and 99.7% of all observations fall, respectively, within plus or minus one, two, and three standard deviations from the mean. Just remember that plus or minus two standard deviations in a bell curve captures about 95% of all observations.
Endogeneity. A circular relationship between X and Y where they feed off each other. It is difficult to disentangle endogenous variables when trying to make a causal argument.
False negative. A situation where statistical tests fail to suggest that something unusual is happening when something is indeed going on. An example is a fire alarm that fails to sound when

a serious fire erupts in a home. False negatives are sometimes called a Type II error or β.

False positive. A situation where statistical tests suggest that something unusual is happening when nothing is afoot. An example is a fire alarm sounding when we burned some toast. False positives are sometimes called a Type I error or α.

Falsification. The act of ruling something out. Outside of pure mathematics, it's difficult to prove anything. The best we can do is to falsify explanations until we are left with one that stands up to repeated challenges.

Fit. How well our model explains relationships among its variables. A high level of fit suggests that our model does a decent job predicting or explaining things; a low level of fit suggests that we should not place too much credence in our model when making sense of the world around us. A basic measure of fit is r^2, where a value closer to +1 is considered better. Propeller heads have many measures of fit.

Geometric mean. A measure of central tendency used to calculate average rate of change for variables that are multiplied together. Often used to measure investment returns over multiple accounting periods, the geometric mean shows the compounded periodic growth rate required for a starting balance to end up equal to the final value in the time series.

Harmonic mean. The average rate for a sequence of steps that have identical effort, such as legs of a relay race. A car that goes uphill for 100 miles and burns 10 gallons of gasoline and then goes downhill for another 100 miles and uses just one gallon of gasoline shows a harmonic mean of 18.2 miles per gallon over the two legs of the journey.

Histogram. A way to show frequency distributions (graphs where the horizontal x-axis shows different values of the variable of interest while the vertical y-axis shows counts for how frequently a particular range of values occurs) for continuous data. Histograms create bins that convert continuous data into discrete data ranges and then count the number of observed values within these buckets. Histograms reveal the presence of multiple modes, asymmetry, and long tails.

Hypothesis [*H*]. A proposed, testable relationship between two or more variables within a model. An example would be the hypothesis that there is a positive relationship between intelligence quotient and annual salary among workers within a given organization. Properly constructed hypotheses may be falsified.

Induction. A form of reasoning that moves from specific examples to generalizations expected to hold outside of the original sample. If we survey five dentists and find that four recommend a brand of toothpaste, then using induction leads us to say that 80% of all dentists recommend that brand.

Inference. The act of drawing conclusions about a population from analysis of a sample drawn from the population. Inference is a form of inductive reasoning, where we reach generalizations from studying specific things.

Inferential statistics. A collection of tools used to draw provisional conclusions about an unobservable population based on statistics associated with an observed sample. An example is estimating a population mean from sample statistics associated with a random sample (e.g., $\mu \approx x \pm 2[s/\sqrt{n}]$ @ 95% confidence level).

Information. A surprise (*Wow, I didn't know that!*) that reduces uncertainty. Numeracy helps us extract information from data.

Interval. A type of scale that has observations separated by uniform differences but that are not measured from an absolute baseline, such as hour of the day. Interval data may be added or subtracted but not multiplied or divided: 4 p.m. is two hours later than 2 p.m. but is not twice as late.

Kurtosis. A measure of the thickness of tails of a distribution relative to a bell curve. High kurtosis suggests fat tails and a greater likelihood of outliers than would be observed in a normal distribution.

Law of large numbers. A relationship linking sample size, sample statistics, and population parameters. As sample size increases, the sample mean converges on the population mean. If the sample is the entire population, then the sample mean equals the population mean.

Lurking variable. An unconsidered third variable that causes correlation in both *X* and *Y*. Correlation between ice cream sales and boat rentals may be due to the lurking variable of sunny weather.

Margin of error. An interval that communicates an assessment of how far from a predicted value that the actual, unobserved number could fall. If we said $\mu = 10 \pm 2$, the margin of error for an unobserved parameter is plus or minus two units from the predicted value of 10. Margins of error are usually associated with confidence levels.

Median. A measure of central tendency calculated by selecting the midpoint of a sorted list of observations. The median cuts the data set into two halves, where 50% of the observations are above and below this measure. If there are an even number of observations, the median is the arithmetic mean of the two observations straddling the midpoint. Median is less influenced by outliers than is arithmetic mean.

Mediation. An intermediate variable (*M*) that may explain why *X* causes *Y*. The presence of rain (*X*) brings about a desire to avoid getting wet (*M*) that causes strollers to open their umbrellas (*Y*). Some people, such as a smitten Gene Kelly in *Singin' in the Rain*, may not care about getting wet and choose to keep their umbrellas closed.[1]

Mode. A measure of central tendency that represents commonly appearing values of observations in the data set. A data set may have one, multiple, or no modes. Modes are helpful when working with dichotomous and discrete variables or nominal and ordinal scales.

Model. A simplified theory that lends itself to testing. Models have variables that operationalize constructs. University of

---

[1]I thank my former statistics professor James Gaskin for providing this example.

Wisconsin professor George Box noted that all models are wrong but some are useful.

Moderation. The presence of a third variable that causes an X variable to have an outsized influence on the Y variable. An example is serving cold milk (moderator) with chocolate cake (X) to amplify a pleasant dessert experience (Y).

Nominal. A form of data that fall into categories that have no logical order (e.g., hair color or make of automobile). Nominal data may be listed alphabetically, in order of how often they occur, or placed in another ordering scheme that makes reference to information outside of the category labels.

Normal distribution. A bell-shaped histogram that results from data pulled from many random, independent trials. Normal distributions, defined by arithmetic means and standard deviations, are constructs used to make predictions. Many data sets are not normally distributed.

Null hypothesis [H_0]. Expressing a prediction in a way that (i) suggests nothing interesting is happening and (ii) the statement may be disproven. An example would be "There is no relationship between eating habits and weight gain." We create and then try to falsify null hypotheses to mitigate confirmation bias.

Numeracy. The craft of extracting information from data and then communicating findings clearly. Numerate people reduce uncertainty.

Occam's razor. The idea that, when choosing among competing models of equal effectiveness, one should favor the one using the fewest assumptions. Numerate people prefer simpler models because they are more understandable. No one expects us to create a theory of everything.

Operationalize. The imperfect process of converting abstract constructs into tangible variables that may subjected to statistical tests. An example would be using finished square feet to define and measure house size.

Ordinal. A type of scale where component observations may be placed into a lowest-to-highest sequence, where the distances between the observations are not clear. Examples include Olympic medal colors or ranks in the military. Be careful when using arithmetic operations on ordinal data.

Outliers. Unusual observations far removed from a data set's other observations. If outliers are the result of measurement errors, they should be excluded from the analysis. If they are the result of valid, reliable measurement, then they contain information and should be included in the analysis and given careful consideration.

***p*-value [p].** A measure of the probability that there is no significant difference between a variable of interest and a critical value (a baseline measure such as 0). Low *p*-values (say, below 5%) suggest that something interesting may be going on but do not prove anything.

Population. The complete data set for a variable of interest, such as the set of all human beings on the planet. It may be prohibitively expensive or time-consuming (or downright impossible) to study a population.

Population mean [μ]. The arithmetic mean of all observations in a population. This parameter is often unobservable and inferred from sample statistics.

Power law distribution [$Y = X^{-a}$]. Also known as a Pareto distribution, this function describes downward sloping histograms where higher values of X appear with diminishing frequency. Extremely high values for X appear more often than would be seen under normal distributions. Power law distributions suggest interdependence among observations.

Proposition. A suggested, unverified relationship between constructs used in a theory. An example would be that there is a positive relationship between intelligence and job performance. Propositions form the basis for hypotheses, which associate variables that operationalize constructs.

Ratio. Data that are measured in terms of equally spaced units from an absolute zero. Observations within ratio data may be added, subtracted, multiplied, and divided. Height, weight, and accounting balances are examples of ratio data. Commonly taught statistical tools assume that observations are composed of ratio data.

Reasoning. The process we use to reach conclusions from raw data. A well-reasoned argument shows the bridge built to link evidence to conclusions.

Reliability. The degree to which successive measurements yield the same result. A reliable scale will return the same reading should someone step on and off the instrument several times without changing anything else.

Reverse causation. Getting our causal story backwards. Even though we argue that X causes Y, what's really happening is that Y causes X.

Sample. When used as a verb, sample means the process of extracting some observations from a population for further study. When used as a noun, sample means the set of observations that have been pulled from the population for detailed analysis.

Sample bias. A problematic situation where observations are extracted from a population in a manner where the observations are not representative of the population. Sample bias brings bad inferences.

Sample size [*n*]. A count of the number of observations in a sample being studied. Thin samples ($n < 30$) may lead to incorrect inferences even if our math is perfect. Very large samples bring small standard errors and the risk of false positives in hypothesis testing.

Skew. The degree of asymmetry associated with a data set. In a normal distribution, high values offset low values so that the shape of the curve to the left of the mean mirrors that on the right, resulting in a skew of 0. A skewed data set, by contrast, has a longer tail that points to one side or the other.

Simpson's Paradox. A situation where a trend disappears or reverses when additional data are included in the analysis. This problem reminds us that all inferences are provisional.
Slope [$r(s_Y/s_X)$]. A measure of how much a second variable (*Y*) moves given change in the first variable (*X*), assuming a linear relationship between the two.
Spurious correlation. A situation where we find a strong linear relationship between two variables when the results are simply due to chance. A way to rule out spurious correlation is to determine whether the relationship continues to hold in a different sample.
Standard deviation [s]. A measure of variance that displays the average distance that observations lie from their arithmetic mean. The standard deviation of a sample is shown with a lowercase s and of a population with the Greek letter σ (sigma).
Standard error [$s/\sqrt{n}$]. A measure of the distance from an observable sample mean (x) to an unobservable population mean (μ). Standard error is used to determine the margin of error associated with a given confidence level.
Statistics. A part of mathematics that evaluates how one may make uncertain inferences about an unobservable population from the study of one or more incomplete samples.
Theory. A framework that combines constructs to explain the past and/or predict the future. Newton's theory of gravity makes predictions based on the constructs of mass and distance.
Trial. Another word for sample when we contemplate pulling repeated samples. Note that a sample may contain one or multiple observations. On a manufacturing line, a quality control engineer could pull a dozen samples, each involving measurement of five units. There were 12 trials involving examination of 60 units.
Validity. The degree to which a variable measures the properties contemplated in a construct. Intelligence quotient (IQ) is probably not a valid measure for the construct of soft skills.

Variable. Something measured in the world around us that serves as a proxy for a construct used in a theory. An example would be using annual salary rate to measure job performance.
Variance. The degree to which observations are far removed from their averages. Variance degrades the usefulness of averages. The formal definition of variance is the square of the standard deviation (i.e., s^2 or σ^2). When used in this book, variance is the superset for measures of dispersion such as standard deviation, skew, kurtosis, coefficient of variation, and volatility.
Volatility. A measure of variance over time. Volatility is often used to assess how much a security price bounces around and provides a measure of investor skittishness.
Zero [0]. An oval-shaped symbol that signifies the absence of something to be measured. Zero helps us measure differences between numbers, increase the precision of measurements, and establish a baseline so that we may calculate ratios.
***z*-score [*z*].** The distance a particular observation strays from a data set's arithmetic mean, where the distance is measured in standard deviations. Values below –3 or above +3 are often classified as outliers.

APPENDIX B – TEN MATH FACTS NUMERATE PEOPLE SHOULD KNOW

Numerate people should be familiar with the following 10 math facts. I have used each of these tools in my career. No need to remember the specifics; just file the conclusions in your mind and look up the particulars should the need arise.

1. **Adding the same positive number to a numerator and a denominator brings the ratio closer to 1.**
   Adding the same positive quantity to the top and bottom halves of a fraction brings the resulting ratio closer to 1. If three students in a sample of five scored a question correctly, the sample pass rate is 60%. If we expand the sample by one student who scored correctly, adding the one to the top and bottom numbers raises the percentage from 60% [3/5] to 67% [(3 + 1)/(5 + 1)], a ratio closer to 1. If we have a ratio of 4.50:1, adding a one to both terms reduces the ratio to 2.75:1.
   This rule matters should, say, an auditor recommend "grossing up" a fraction by a fixed sum. Users of financial

statements often compute the ratio of current assets to current liabilities to assess a firm's ability to pay short-term obligations. An audit adjustment that separates a net asset into a distinct asset and offsetting borrowing may not be welcomed by the firm's controller: adding the same balance to balance sheet components showing strong current assets relative to current liabilities will lower the current ratio closer to 1 and make a healthy firm appear somewhat less solvent.

2. **Dividing a growth rate into 72 gives the number of periods for a quantity to double.**
This formula allows someone to evaluate growth rates using mental math. The rule of 72 permits comparisons of growth rates and time periods needed to double quantities. For example, if a country's annual population growth rate is 2%, then we can expect the number of residents to double in about 36 years (72/2). If the growth rate rises to 3%, the interval will shrink to about 24 years (72/3).
The math behind this rule rests on the association between a growth rate (g) and the number of time periods (t). For a quantity to double (say, going from 1 to 2), we use the following equation: $(1 + g)^t = 2$. The rub is that we have two unknowns and one equation. Here is a solution that meets our needs.

Take the natural logarithm of both sides:	$\ln(1 + g)^t = \ln(2)$
Use the exponents rule to bring down the variable t:	$t[\ln(1 + g)] = 0.69$
Divide both sides by the expression $\ln(1 + g)$:	$t = 0.69/\ln(1 + g)$
Assume that g is small, so then $\ln(1 + g) \approx g$:	$t = 0.69/g$
Substitute the number 0.72 for 0.69:	$t = 0.72/g$

The number 72, divisible by many integers, makes mental math simple. We may also flip this equation to estimate a required growth rate if we have a required number of time periods for a quantity to double. For example, if a firm's CEO promises that revenue will double in six years, then the firm must achieve a compound annual growth rate of about 12% (72/6) over that period.

3. **A transposition error is divisible by 9.**
When two adjacent numbers are transposed – put in the wrong order, such as 1,243 inadvertently representing the intended balance of 1,234 – the difference in the resulting sum is divisible by 9. Some people invert digits when copying down figures. For example, consider the following table, in which one entry has a transposition:

Correct Values		*One Transposed Value*		*Difference*	
5,436		5,436			
235		235			
8,744		8,744			
3,296	⟷	2,396			
1,572		1,572			
67		67			
1,485		1,485			
7,587		7,587			
28,422		27,522	=	900	Difference is divisible by 9.

Proofs of why this relationship holds are available on the Internet. This trick is useful when looking for a problem in reconciling a checkbook to a bank statement. In my experience, the transposition results from a recording error from the check writer, not the bank!

4. **Multiplied quantities are more likely to start with 1 or 2 than an 8 or 9.**
Leading significant digits (numbers other than zero) of naturally occurring quantities – measures of things like

accounting balances or population sizes that span orders of magnitude – tend to be small. Quantities following Benford's law have the following distributions of leading digits:

Leading Digit	*Relative Frequency*
1	30.1%
2	17.6%
3	12.5%
4	9.7%
5	7.9%
6	6.7%
7	5.8%
8	5.1%
9	4.6%
	100.0%

To take an example, consider Minnesota, the so-called land of 10,000 lakes. Suppose that figure is accurate and we wish to estimate the distribution of lake surface areas, as measured in square meters. Using Benford's law leads us to predict that about 3,000 of these lakes (30.1% × 10,000) have surface areas that begin with the number 1 (e.g., 15, 142, or 1,876 m^2) and that fewer than 500 start with 9.

Benford's law is helpful when looking for accounting fraud. A distribution of invoices or checks with an unexpectedly high number of transactions starting with large leading digits is evidence of the presence of artificial quantities (Durtschi et al., 2004).

5. **For a sample size of n, the margin of error in an election poll is $\pm 1/\sqrt{n}$.**

When elections approach, pollsters make predictions about the fraction of participating voters who are expected to choose a candidate. Such predictions are subject to a margin of error, discussed in Chapter 10, except that in this case the population parameter is a ratio (p) instead of an average (x).

If the pollster has avoided a biased sample (which is not easy), then we may determine the margin of error given the sample size, or vice versa. At a 95% confidence level, the margin is the reciprocal of the square root of the sample size ($1/\sqrt{n}$). The derivation uses the expression $\sqrt{[(p)(1-p)/n]}$ as the measure of standard error for p, a value between 0 and 100%.

The margin of error for a proportion at a 95% confidence level:	$\pm 2\sqrt{[(p(1-p))/n]}$
Assume a close race so that p and $1-p$ are equal to 50%:	$\pm 2\sqrt{[(0.5 \times 0.5)/n]}$
Multiply p times $(1-p)$:	$\pm 2\sqrt{[0.25/n]}$
Pull out the square root of 0.25:	$\pm 2(1/2)\sqrt{[1/n]}$
Multiply 2 times 1/2:	$\pm 1/\sqrt{n}$

If the sample size is 1,000 voters, then the margin of error at a 95% confidence level would be $\pm\ 1/\sqrt{1{,}000} = \pm 3.2\%$. If the pollster estimates p at 57%, then, assuming nothing changes between the time of polling and the election, there is at least a 95% chance that the candidate will secure more than 50% of the votes.

We can also work backwards and infer the sample size, if the margin of error is given. A margin of error of ±5% implies a sample size of 400 (if $0.05 = 1/\sqrt{n}$, then $\sqrt{n} = 20$ and $n = 400$).

6. **There is a 94% chance that the median is bounded by the high and low figure in a random sample of five observations.** The so-called "rule of five" (Hubbard, 2007, p. 29) says that there is a 94% chance that the median of a population falls

within the range covered by the smallest and largest values among a random sample of five items. Thus, if a random sample of observations had the values 5, 8, 11, 12, and 14, then we may infer with a 94% level of confidence that the population median is between 5 and 14.

The math behind this idea is that there is a 50% chance that any given observation will fall below the population median, *M*. With independent samples, there is a 3.125% chance (50% raised to the fifth power) that all five samples will be smaller than *M*. Similarly, there is a 3.125% chance that all five samples will be greater than *M*. The combined probability that all samples are bigger or smaller than *M* is 3.125% + 3.125%. Thus, the chance that at least one observation is smaller than *M* and at least one observation is larger than *M* is 100% – 6.25% ≈ 94%.

This rule is useful is when we know nothing about central tendency within a population. A sample of five observations can help us reduce uncertainty quickly and inexpensively.

7. **The sum of digits equals the product of $(n + 1)(n/2)$.**

Suppose a coach asks you to perform a sequence of exercises where each interval involves one fewer repetition. For example, the coach tells you to perform 10 pushups, then nine, then eight, and so on. As your arms burn, you wonder how many pushups you must endure.

The sum of digits calculation provides an easy answer. To calculate the sum of all numbers from 1 to a terminal value, *n*, simply add 1 to *n* and then multiply this sum by one-half of *n*. So, if one wanted to add the numbers from 1 to 10, the quick answer would be 11 × 5 or 55.

The scope of Gauss's genius, mentioned in Chapter 6, was allegedly revealed when a grade school teacher asked him to add up all the digits from 1 to 100, and the prodigy immediately returned the answer of 5,050. Gauss's approach was to frame the problem as the sum of sums. He reasoned that 1 plus 100 equals 101, 2 plus 99 also equals 101, 3 plus 98 again equals 101, and so on. Thus, the sum of all the digits equals 101 multiplied by the number of pairs in the series, in this case 50.

8. **In a room with 23 people, there is a 50% chance that two share the same birthday.**

The so-called birthday problem illustrates the prevalence of seemingly unlikely events. The trick is to look at the complement of probabilities. If the percentage chance of something happening is p, then the likelihood of that event not happening is $1 - p$.

Assume that birthdays are evenly distributed across a 365-day year. A person walks into a room. Obviously, there is no one else present who shares his or her birthday. When a second person enters the room, there is a small chance that these two were born on the same day. Expressed as a ratio, there is a 1-in-365 (0.3%) chance that the second person shares the birthday of the first person. Expressed as a complement, there is a 99.7% (1 – 0.3%) chance that the two were not born on the same day.

Things get trickier when a third person enters the room. A unique birthday requires the new arrival to have a birth date that *avoids* the birthdays of the first two people. The pool of available unique birthdays falls with each arrival. While there are still 365 days in a year, the number of remaining unused birthdays has been reduced by two to 363. The chance of the third person having a distinctive birthday is 363/365.

The likelihood of two independent events is the product of the individual probabilities. For example, the probability of getting two heads in two successive coin flips is one-half times one-half. Here, the probability that both arrivals avoid their predecessors' birthdays is the product of (364/365) × (363/365) = 99.7% × 99.5% = 99.2%. Building on this line of reasoning, if 23 people enter the room, then the probability that each has avoided everyone else's birthday is the following scary chain of calculations:

2	3	4	5	6	7	8	9	10	11	12	13	14	15	16	17	18	19	20	21	22	23
$\frac{364}{365}\times$	$\frac{363}{365}\times$	$\frac{362}{365}\times$	$\frac{361}{365}\times$	$\frac{360}{365}\times$	$\frac{359}{365}\times$	$\frac{358}{365}\times$	$\frac{357}{365}\times$	$\frac{356}{365}\times$	$\frac{355}{365}\times$	$\frac{354}{365}\times$	$\frac{353}{365}\times$	$\frac{352}{365}\times$	$\frac{351}{365}\times$	$\frac{350}{365}\times$	$\frac{349}{365}\times$	$\frac{348}{365}\times$	$\frac{347}{365}\times$	$\frac{346}{365}\times$	$\frac{345}{365}\times$	$\frac{344}{365}\times$	$\frac{343}{365}$

The product of these ratios is 49.3%. The complement of this fraction, showing the chance that we violate the no-sharing rule, is 1 – 49.3% or 50.7%. Put simply, there is more than a 1-in-2 chance that 23 people in a room will share a birthday.

This relationship allows for an easy bar trick. The next time you're in a crowded bar, bet your buddy that two people present share the same birthday. Added to this statistical logic is the fact that many people go out to celebrate birthdays with cocktails. The cost of the wager is that you'll have to poll the room, but odds are that you'll earn a free drink in addition to securing some serious bragging rights.

9. **Pizza joints offering three toppings among 15 choices have 455 possibilities.**

   This is an example of an "*n* choose *k*" statistics problem, where *n* is the number of possibilities among things to be selected and *k* is the constraint for how many possibilities may be selected for a given trial. With the pizza joint, *n* is the 15 possible toppings (e.g., onions, green peppers, black olives, pepperoni, mushrooms, bacon, and so on) and *k* is the limit of three toppings that may be selected at a given price for the pizza.

   The formula is $n!/(k![n - k]!)$, where the exclamation mark means factorial. The factorial for 15 equals this number multiplied by every positive integer below it:

$$15! = 15\times14\times13\times12\times11\times10\times9\times8\times7\times6\times5\times4\times3\times2\times1 = 1,307,674,380,000$$

   In our example, 15 choose 3 is calculated as $15!/(3! \times [15 - 3]!) = 455$.

Online calculators will do this work for you. Other labels for this formula are nCk or the binomial coefficient.

10. **Perhaps one-third of reported laboratory discoveries are wrong.**

In Chapter 13, we showed how false positives (also known as α) and false negatives (sometimes signified as β) complicate our ability to properly reject null hypotheses. Considering both types of errors together shows why we should be cautious when accepting findings from experiments.

Some researchers are willing to live with an α of 5% (i.e., a 1-in-20 chance that statistical tests suggest that something is going on when nothing is really happening) and β of 20% (a 1-in-5 chance that statistical tests fail to detect something that is really there).

Consider the following chart. As fallible humans, we almost never know whether something is True with a capital T. Instead, we resort to imperfect statistical tests that help us make sense of the hazy world in which we live. In the chart, our line of sight is limited to the two horizontal rows, in which imperfect statistical tests do or do not detect significant findings. The ability to look down the vertical columns to determine whether a hypothesis is true or false is reserved for supreme beings with powers of observation we will never have.

		Unobservable Reality	
		Hypothesis is False	Hypothesis is True
		900	100
Statistical Conclusions	Nothing detected	855	20 β = 20%
	Something detected	45 α = 5%	80

Let's assume a lab investigates 1,000 hypotheses, where 10% are true. Then, 900 of the posited hypotheses are wrong. Unfortunately, with $\alpha = 5\%$, 45 of these 900 hypotheses show false positives. Further, since $\beta = 20\%$, only 80 of the 100 true hypotheses show statistical results that detect something is happening. The combination of Type 1 and Type II errors brings about lab results such that a third of the cases where statistical tests reveal something interesting is happening when in fact nothing is going on (45/[45 + 80] = 36%).

Lab results are likely worse than this because journal editors favor manuscripts showing the presence of statistical significance. I believe that some researchers, under pressure to publish, stretch to see things that are not there, raising α and aggravating this problem.[1]

[1]A good discussion of this phenomenon is presented in "Trouble at the Lab," *Economist*, October 19, 2013.

REFERENCES

Andriani, P., & McKelvey, B. (2007). Beyond Gaussian averages: Redirecting international business and management research toward extreme events and power laws. *Journal of International Business Studies, 38*(7), 1212–1230. http://www.palgrave-journals.com/rpm/free_articles_introduction.html

Bernstein, P. (1996). *Against the gods: The remarkable story of risk.* Wiley.

Bickel, P., Hammel, E., & O'Connell, J. (1975). Sex bias in graduate admissions: Data from Berkeley. *Science, 187*(4175), 398–404. https://doi.org/10.1126/science.187.4175.398

Bierce, A. (1911). *The devil's dictionary.* The World Publishing Company.

Bookstaber, R. (2017). *The end of theory: Financial crises, the failure of economics, and the sweep of human interaction.* Princeton University Press.

Diaconis, P., Holmes, S., & Montgomery, R. (2007). Dynamical bias in the coin toss. *SIAM Review, 49*(2), 211–235. https://doi.org/10.1137/S0036144504446436

Durtschi, C., Hillison, W., & Pacini, C. (2004). The effective use of Benford's law to assist in detecting fraud in accounting data. *Journal of Forensic Accounting, V*(1), 17–34.

Ehrenberg, A. (1977). Rudiments of numeracy. *Journal of the Royal Statistical Society (Series A), 140*(3), 277–297.

Forer, B. (1949). The fallacy of personal validation: A classroom demonstration of gullibility. *Journal of Abnormal and Social Psychology, 44*(1), 118–123. https://doi.org/10.1037/h0059240

Friedman, M. (1953). *Essays in positive economics.* University of Chicago Press.

Gladwell, M. (2008). *Outliers: The story of success.* Little, Brown and Company.

Gould, S. (2013). The median isn't the message. *AMA Journal of Ethics, 15*(1), 77–81.

Hagerty, J. (2016, December 30). Vera Rubin forced the cosmological theorists to think again. *Wall Street Journal.* https://www.wsj.com/articles/vera-rubin-forced-the-cosmological-theorists-to-think-again-1483110002

Hawkins, J., & Blakeslee, S. (2004). *On intelligence.* Times Books.

Hubbard, D. (2007). *How to measure anything: Finding the value of "intangibles" in business.* Wiley.

Hudson, P. (2000). *History by numbers.* Arnold.

King, T. (2017). The problem with non-GAAP earnings. *Strategic Finance, 98*(5), 31–39.

King, T. (2018). The observer effect and U.S. accounting rules. *Engaged Management ReView, 2*(2), 30–38.

Knight, F. (1921). *Risk, uncertainty and profit.* Houghton Mifflin Company.

Markarian, J. (2018, May 15). Who needs calculus? Not high-schoolers. *Wall Street Journal.*

Menand, L. (2001). *The metaphysical club: A story of ideas in America.* Farrar, Straus and Giroux.

O'Boyle, E., & Aguinis, H. (2012). The best and the rest: Revisiting the norm of normality of individual performance. *Personnel Psychology, 65*(1), 79–119. https://doi.org/10.1111/j.1744-6570.2011.01239.x

Popper, K. (1959). *The logic of scientific discovery.* Hutchinson & Co.

Rose, T. (2016). *The end of average: How to succeed in a world that values sameness.* HarperOne.

Rosling, H., Rosling, O., & Rönnlund, A. R. (2018). *Factfulness: Ten reasons why we're wrong about the world – and why things are better than you think.* St. Martin's Press.

Rubin, R. (2009, January 9). Citigroup statement on Rubin's departure; Rubin's letter. *Wall Street Journal.*

Ruggles, R., & Brodie, H. (1947). An empirical approach to economic intelligence in World War II. *Journal of the American Statistical Association, 42*(237), 72–91.

Russell, B. (1912). *The problems of philosophy.* Henry Holt and Company.

Salsburg, D. (2001). *The lady tasting tea: How statistics revolutionized science in the twentieth century.* Henry Holt and Company.

Shannon, C. (1948). A mathematical theory of communication. *The Bell System Technical Journal, 27*(3), 379–423. https://doi.org/https://doi.org/10.1007/978-3-662-05322-5_10

Stevens, S. (1946). On the theory of scales of measurement. *Science, 103*(2684), 677–680.

Stigler, S. (1986). *The history of statistics: The measurement of uncertainty.* Harvard University Press.

Stigler, S. (1999). *Statistics on the table: The history of statistical concepts and methods.* Harvard University Press.

U.S. Securities and Exchange Commission. (2015, August 5). SEC adopts rule for pay ratio disclosure. Press release 2015-160. https://www.sec.gov/news/pressrelease/2015-160.html.

Weick, K. (1995). *Sensemaking in organizations.* SAGE Publications.

Ziliak, S. (2019). How large are your G-values? Try Gosset's Guinnessometrics when a little "p" is not enough. *American Statistician, 73*(suppl. 1), 281–290. https://doi.org/10.1080/00031305.2018.1514325

Ziliak, S., & McCloskey, D. (2008). *The cult of statistical significance: How the standard error costs us jobs, justice, and lives.* University of Michigan Press.

Ziporyn, T. (1992). *Nameless diseases.* Rutgers University Press.

ABOUT THE AUTHOR

Tom King is chair of the Department of Accountancy at Case Western Reserve University's Weatherhead School of Management, having twice earned the School's Graduate Teaching Excellence Award.

King studies how information helps people within organization pursue common goals. His previous book is *More Than a Numbers Game: A Brief History of Accounting.*

He spent 30 years at Progressive Insurance, working in various line management and finance roles. He was general manager of the business unit that sold the first-ever car insurance policy purchased by a motorist over the Internet and head of investor relations when Progressive became the first company in the world to disclose financial statements on a monthly basis. Before joining Progressive, he worked on the audit staff of Arthur Andersen & Co. in New York.

Married with three grown children, King lives in Chagrin Falls, Ohio. He enjoys crime novels, art museums, and the Cleveland Metroparks. He earned degrees from Harvard College,

New York University, Harvard Business School, and CWRU, and holds CPA and CMA certifications. Management education, he believes, should rest on a liberal arts foundation. This is the book he wishes someone had handed him when he began his career.

INDEX